Continued Fractions

CONTINUED FRACTIONS

Andrew M Rockett

Department of Mathematics
Long Island University
C. W. Post Campus

Peter Szüsz

Department of Mathematics
State University of New York
Stony Brook

World Scientific
Singapore • New Jersey • London • Hong Kong

Published by

World Scientific Publishing Co. Pte. Ltd.
P O Box 128, Farrer Road, Singapore 9128
USA office: Suite 1B, 1060 Main Street, River Edge, NJ 07661
UK office: 73 Lynton Mead, Totteridge, London N20 8DH

First published 1992
First reprint 1994

CONTINUED FRACTIONS

ISBN 981-02-1047-7

Printed in Singapore by Utopia Press.

CONTENTS

PREFACE

The theory of continued fractions does not receive the attention it deserves. While the standard textbooks on number theory do present the most basic properties of regular continued fractions, no real use is made of the power contained within those simple properties. Worse still is the practice of some authors to conclude their discussions of continued fractions with either a demonstration of a theorem of Hurwitz (on rational approximations to a given irrational number) or a discussion of Pell's equation $x^2 - dy^2 = 1$ as if it were a culminating application of the subject.

Some books devoted to continued factions are available. The comprehensive classic of Perron [1954] treats both the arithmetic and analytic properties of many kinds of continued fractions and is a revision of earlier (1913 and then 1929) editions. But it is the 1929 version that is available in the United States as a reprint from the Chelsea Publishing Company. There are two works dealing only with the analytic theory: Wall [1973], which is a reprint of a book written in 1948, and a more recent presentation by Jones and Thron [1980]. Last, but from our viewpoint not the least, is the beautiful compendium by Khintchine [1963] on the arithmetic and metrical properties of regular continued fractions. The fact that this book has gone through three Russian editions since 1935 and several translations (into German and English) is a clear indication of the continuing interest in and the importance of its contents, to say nothing of the vitality of the exposition. However, the reissue of such a classic from time to time neither guarantees that the results remain the best possible nor insures that the proofs are the simplest and most direct. Our purpose with the present book is to attempt this revision and to present some applications of the theory of continued fractions to problems from number theory, the geometry of numbers, and diophantine approximation.

While our presentation is self-contained, we assume that the reader is familiar with the basic notions of number theory (see Hardy and Wright [1971], LeVeque [1977], or Niven and Zuckerman [1980]) and has a rudimentary acquaintance with Lebesgue measure theory (such as used in Niven [1956] or Kac [1959]). Frequently but implicitly, we shall make use of the well-known Lebesgue dominated convergence and the Beppo-Levi theorems discussed in introductions to the Lebesgue integral. Since we find it most natural to express the metrical properties of continued fractions in the language of probability, we also develop the required terminologies and tools from that theory (see Rényi [1970]) and then the principal results of Chapter V become strong laws of large numbers (see also Révész [1968] and Kac [1959] for additional discussions of laws of large numbers). Lévy [1937] concluded his book on sums of random variables with a discussion of continued fractions and his proofs for some of the results in Chapter V. Part of the metrical results, such as Khintchine's theorem and its applications, may also be deduced from ergodic theory (see Billingsley [1965], Kac [1959] and Ryll-Nardzewski [1951] for such treatments). We have avoided further mention of the ergodic theorem and its applications both for simplicity and our need for explicit error estimates in Chapter VI. We derive precise convergence rates from our proof of the Gauss-Kuzmin theorem, which allow us to show that the partial quotients of regular continued fractions are "weakly dependent" random variables (it also is possible to obtain central limit theorems and laws of the iterated logarithm for such quantities; see Philipp [1971]). Further information on the historical aspects of continued fractions prior to 1900 may be found in Wölffing [1908]. Readers interested in the historical context of the classical results are referred to Weil [1984].

We thank Professor Bodo Volkmann, for his careful reading of our manuscript, and Alexander Durner, for his useful comments after using an earlier draft as the basis for an undergraduate seminar at the University of Stuttgart during the 1989-90 winter semester.

NOTATIONS

Equations and results (theorems and lemmas) are numbered consecutively within each section and are identified by their chapter, section and number; thus "Theorem II.3.1" refers to Theorem 1 of §3 in Chapter II. References to items outside the section but within the same chapter omit the chapter designation and references within the section consist of just the number.

A citation to an item in the bibliography is made by stating the author and the date of publication. Items having the same author and year of publication are distinguished by adjoining lowercase Roman letters to the date; for example, "Descombes [1956b]." Most citations are given in the NOTES at the end of each chapter.

$a \mid b$ means that the integer a divides the integer b; that is, there is an integer k such $a \cdot k = b$. a is congruent to b modulo m, written $a \equiv b \bmod m$, if $m \mid (a - b)$. The greatest common divisor of two integers a and b is denoted by (a, b). We assume that the reader is familiar with the calculation of (a, b) via the Euclidean algorithm and the representation of (a, b) by the expression $ax + by$, where x and y are integers.

For a given real number x, we write $[x]$ for the greatest integer not exceeding x, $\{x\}$ for the fractional part of x and $\| x \|$ for the distance from x to the nearest integer. $[x, y]$ denotes the closed interval of real numbers t such that $x \leq t \leq y$ and we write $|[x, y]|$ for the length $y - x$ of this interval.

To compare the asymptotic behavior of two functions, we use the Landau "order notation" $f(x) = \mathcal{O}(g(x))$ to indicate that there exists a constant K independent of x such that $|f(x)/g(x)| < K$ for x sufficiently large. Thus, for example, $x^n = \mathcal{O}(x^{n+1})$ for any positive integer n, and if $f_k(x) = \mathcal{O}(q^k)$ for $k = 1, 2, \ldots$ with the same constant K and $0 < q < 1$ is a constant, then $\sum f_k(x) = \mathcal{O}(1)$.

Chapter I

INTRODUCTION

§1. What is a continued fraction?

The theory of continued fractions arises from consideration of expressions given in the form

$$a_0 + \cfrac{1}{a_1 + \cfrac{1}{a_2 + \cfrac{1}{\cdots + \cfrac{1}{a_n}}}} \tag{1}$$

whence the name *continued fraction*. The quantities a_0, a_1, a_2, ..., a_n are called the *partial quotients* and may be taken to be integers, real or complex numbers, or functions of such variables. Further, more complicated continued fractions may be considered in which the numerators are no longer positive ones.

The expression (1) can be brought into the form of a rational function of the partial quotients. If we begin by considering just the first few terms of (1), we may write

$$C_0 = a_0,$$

$$C_1 = a_0 + \tfrac{1}{a_1} = \frac{a_0 a_1 + 1}{a_1},$$

$$C_2 = a_0 + \cfrac{1}{a_1 + \tfrac{1}{a_2}} = \frac{a_0(a_1 a_2 + 1) + a_2}{a_1 a_2 + 1},$$

and so on. These C_k's are the *convergents* of the continued fraction and it is clear that the form of C_{k+1} depends on the previous C_k's and a_{k+1}.

Let us define two sequences $\{A_k\}_{k \geq -1}$ and $\{B_k\}_{k \geq -1}$ in terms of the a_k's by setting

$$A_{-1} = 1, A_0 = a_0 \quad \text{and} \quad B_{-1} = 0, B_0 = 1$$

$$\begin{cases} A_{k+1} = a_{k+1}A_k + A_{k-1} \\ B_{k+1} = a_{k+1}B_k + B_{k-1} \end{cases} \quad \text{for } 0 \leq k \leq n-1. \tag{2}$$

We notice immediately that $A_0/B_0 = a_0/1 = C_0$ and $A_1/B_1 = (a_1 a_0 + 1)/a_1 = C_1$, and we claim that the fraction A_k/B_k is precisely the convergent C_k. Suppose that this claim is established for convergents of order less than or equal to k. Then for an expression with $k+1$ terms we may rewrite it as an expression in just k terms by making the new last term $a_k + 1/a_{k+1}$. But the k-th convergent of our new expression is

$$\frac{\left(a_k + \frac{1}{a_{k+1}}\right)A_{k-1} + A_{k-2}}{\left(a_k + \frac{1}{a_{k+1}}\right)B_{k-1} + B_{k-2}}$$

and the quantities A_{k-1}, A_{k-2}, B_{k-1} and B_{k-2} coincide with those of our original continued fraction. This convergent is the same as the C_{k+1} we wished to determine, so we have that

$$C_{k+1} = \frac{(a_k A_{k-1} + A_{k-2}) + (\frac{1}{a_{k+1}})A_{k-1}}{(a_k B_{k-1} + B_{k-2}) + (\frac{1}{a_{k+1}})B_{k-1}}$$

$$= \frac{a_{k+1}A_k + A_{k-1}}{a_{k+1}B_k + B_{k-1}} = \frac{A_{k+1}}{B_{k+1}},$$

as required. The number A_k is the *numerator* and B_k is the *denominator* of the convergent.

In addition to forming the convergents, the A_k's and B_k's also have the property that

$$A_k B_{k-1} - A_{k-1} B_k = (-1)^{k-1} \quad \text{for } k \geq 0 \tag{3}$$

since $A_0 B_{-1} - A_{-1} B_0 = 0 - 1$ and

$$A_{k+1}B_k - A_k B_{k+1} = (a_{k+1}A_k + A_{k-1})B_k - A_k(a_{k+1}B_k + B_{k-1})$$

$$= -(A_k B_{k-1} - A_{k-1} B_k) \qquad \text{for } k > 0.$$

§2. Regular continued fractions.

We now specialize the continued fractions (1.1) described in the previous section to *regular continued fractions* in which the initial term a_0 is an integer and the (possibly infinite) sequence $a_1, a_2, \ldots$ of partial quotients consists of positive integers. For the remainder of this book, we shall mean "regular continued fraction" whenever we write "continued fraction" unless we specifically state otherwise. The sequences (1.2) now consist of integers, and the positive integers $\{B_k\}_{k \geq 0}$ are increasing monotonically for $k > 0$. Since we can equate the fraction A_k/B_k with the convergent C_k, we shall dispense with the latter notation and we will refer to the fractions $\{A_k/B_k\}_{k \geq -1}$ as the convergents. Since (1.3) holds, the linear diophantine equation

$$A_k B_{k-1} - A_{k-1} B_k = (-1)^{k-1}$$

shows that the convergents are reduced fractions since $(A_k, B_k) = 1$; we also have that $(A_k, A_{k+1}) = 1$ and $(B_k, B_{k+1}) = 1$ for $k > 0$. Moreover,

$$\frac{A_k}{B_k} - \frac{A_{k-1}}{B_{k-1}} = \frac{(-1)^{k-1}}{B_k B_{k-1}}. \tag{1}$$

Thus

$$\frac{A_0}{B_0} + \sum_{k=1}^{\infty} \left(\frac{A_k}{B_k} - \frac{A_{k-1}}{B_{k-1}} \right) = \frac{A_0}{B_0} + \sum_{k=1}^{\infty} \frac{(-1)^{k-1}}{B_k B_{k-1}}$$

is a convergent series and the limit

$$t = \lim_{k \to \infty} \frac{A_k}{B_k} \tag{2}$$

exists. By (1), the convergents $\{A_k/B_k\}_{k \geq 0}$ approximate t alternately from below and above.

If the continued fraction of t terminates as in (1.1), then $t = A_n/B_n$ is a rational number. If $a_n > 1$, then the same value for t can be obtained from a continued fraction with one more partial quotient by setting $a_{n+1} = 1$ and lowering the value of a_n by one. On the other hand, if $a_n = 1$, the continued fraction for t may be shortened by dropping the last term and increasing a_{n-1} by one. We could avoid this ambiguity by requiring that every terminating regular continued fraction ends with $a_n \geq 2$, but for some purposes it will be convenient to make use of it. If the continued fraction for t does not terminate, then t must be an irrational number because if t were rational, say $t = p/q$ with $(p, q) = 1$, then the positive integers $|B_k p - A_k q| = B_k q |p/q - A_k/B_k|$ would be less than q/B_{k+1} by (1), yet this is less than 1 if k is sufficiently large.

Let us investigate the relationship between t and the convergents $\{A_k/B_k\}_{k \geq 0}$ in more detail. By the *k-th complete quotient* ζ_k of our continued fraction, we mean the continued fraction

$$\zeta_k = a_k + \cfrac{1}{a_{k+1} + \cfrac{1}{a_{k+2} + \cdots}},$$

which we shall write in the more compact form

$$\zeta_k = [a_k; a_{k+1}, a_{k+2}, \ldots]. \tag{3}$$

Of course, $t = [a_0; a_1, a_2, \ldots] = \zeta_0$ and we notice immediately that

$$\zeta_k = a_k + \frac{1}{\zeta_{k+1}} = [a_k; \zeta_{k+1}]. \tag{4}$$

To see that t has the form

$$t = \frac{A_k \zeta_{k+1} + A_{k-1}}{B_k \zeta_{k+1} + B_{k-1}} \quad \text{for } k \geq 0, \tag{5}$$

we check first that

$$t = \frac{A_0 \zeta_1 + A_{-1}}{B_0 \zeta_1 + B_{-1}} = \frac{a_0 \zeta_1 + 1}{\zeta_1} = a_0 + \frac{1}{\zeta_1} = \zeta_0$$

and then verify the induction step by observing that

$$t = \frac{\left(a_{k+1} + \frac{1}{\zeta_{k+2}}\right) A_k + A_{k-1}}{\left(a_{k+1} + \frac{1}{\zeta_{k+2}}\right) B_k + B_{k-1}}$$

$$= \frac{(a_{k+1}A_k + A_{k-1}) + (\frac{1}{\zeta_{k+2}})A_k}{(a_{k+1}B_k + B_{k-1}) + (\frac{1}{\zeta_{k+2}})B_k} = \frac{A_{k+1}\zeta_{k+2} + A_k}{B_{k+1}\zeta_{k+2} + B_k}.$$

Formula (5) will play an essential role in our study of continued fractions. We now show that the form of (5) together with the relation (1.3) is sufficient to define the convergents of t in the following sense.

Theorem 1. *Let* $t = (A\zeta + A')/(B\zeta + B')$ *where* $\zeta > 1$ *and the integers* A, A', B *and* B' *are such that* $B > B' > 0$ *and* $AB' - A'B = \pm 1$. *Then there is an index* k *such that* $A = A_k$, $A' = A_{k-1}$, $B = B_k$, $B' = B_{k-1}$ *and* $\zeta = \zeta_{k+1}$ *in the continued fraction for* t.

Proof. Since $(A,B) = 1$, the finite continued fraction $[a_0; a_1, \ldots, a_k] = A_k/B_k$ for the rational number A/B has $A = A_k$ and $B = B_k$. By the ambiguity mentioned earlier, we may allow $a_k = 1$ if necessary and so choose k to be such that

$$AB' - A'B = (-1)^{k-1}.$$

But now

$$A_k B' - A' B_k = A_k B_{k-1} - A_{k-1} B_k$$

and hence

$$A_k(B' - B_{k-1}) = B_k(A' - A_{k-1}).$$

Since A_k and B_k are relatively prime, B_k must divide $|B' - B_{k-1}|$. But

$$|B' - B_{k-1}| < B_k - B_{k-1} \leq B_k$$

and so $|B' - B_{k-1}|$ must be zero. Thus $B' = B_{k-1}$ and then $A' = A_{k-1}$. Now we have that

$$t = \frac{A_k\zeta + A_{k-1}}{B_k\zeta + B_{k-1}}$$

and also that t is as given by (5). Equating these two expressions, $\zeta = \zeta_{k+1}$ and the proof is finished.

This result essentially says that the number t and any of the "tails" of its continued fraction are related by a "unimodular" transformation. In order to extend this notion to include the fact that ζ_{k+1} and t also have identical "tails," let us say that two real numbers s and t are *equivalent*, $s \sim t$, if they are related by a unimodular transformation:

$$ s = \frac{at+b}{ct+d} $$

where the integers a, b, c and d are such that $ad - bc = \pm 1$ (c and d are no longer restricted as were B and B' in Theorem 1). This relation clearly is reflexive, symmetric and transitive, and any two rational numbers are equivalent (since they are each equivalent to zero) even though their continued fractions are finite and lack "tails." We now show that any two equivalent irrational numbers have identical "tails."

Theorem 2. *Two irrational numbers s and t with continued fractions $s = [a_0';\ a_1',$ $\ldots,\ \zeta_{k+1}']$ and $t = [a_0;\ a_1, \ldots, \zeta_{k+1}]$ are equivalent if and only if $\zeta_{m+1}' = \zeta_{n+1}$ for some indices m and n.*

Proof. Suppose first that $s \sim t$. By an appropriate sign change we may suppose that $s = (at+b)/(ct+d)$ with $ct + d > 0$. Using the representation (5) for t,

$$ s = \frac{a\left(\dfrac{A_k \zeta_{k+1} + A_{k-1}}{B_k \zeta_{k+1} + B_{k-1}}\right) + b}{c\left(\dfrac{A_k \zeta_{k+1} + A_{k-1}}{B_k \zeta_{k+1} + B_{k-1}}\right) + d} = \frac{\left(\left(a\dfrac{A_k}{B_k} + b\right)B_k\right)\zeta_{k+1} + \left(\left(a\dfrac{A_{k-1}}{B_{k-1}} + b\right)B_{k-1}\right)}{\left(\left(c\dfrac{A_k}{B_k} + d\right)B_k\right)\zeta_{k+1} + \left(\left(c\dfrac{A_{k-1}}{B_{k-1}} + d\right)B_{k-1}\right)}. $$

Since $A_k/B_k \to t$ as k increases, the requirements of Theorem 1 are fulfilled provided that k is sufficiently large, since $(ct + d)B_k > (ct + d)B_{k-1} > 0$. Thus $\zeta_{m+1}' = \zeta_{k+1}$ for some appropriate index m.

Conversely, suppose that $\zeta_{m+1}' = \zeta_{n+1}$ for some m and n. Letting ζ represent this common value, (5) gives

$$ t = [a_0;\ a_1, \ldots, a_n, \zeta] = \frac{A_n \zeta + A_{n-1}}{B_n \zeta + B_{n-1}} $$

and

$$s = [a_0'; a_1', \ldots, a_m', \zeta] = \frac{\mathcal{A}_m \zeta + \mathcal{A}_{m-1}}{\mathcal{B}_m \zeta + \mathcal{B}_{m-1}}.$$

Solving for the common value,

$$\zeta = -\frac{B_{n-1}t - A_{n-1}}{B_n t - A_n} \quad \text{and} \quad \zeta = -\frac{\mathcal{B}_{m-1}s - \mathcal{A}_{m-1}}{\mathcal{B}_m s - \mathcal{A}_m}$$

and then

$$\frac{\mathcal{B}_{m-1}s - \mathcal{A}_{m-1}}{\mathcal{B}_m s - \mathcal{A}_m} = \frac{B_{n-1}t - A_{n-1}}{B_n t - A_n}$$

gives

$$s = \frac{(B_{n-1}\mathcal{A}_m - B_n\mathcal{A}_{m-1})t + (A_n\mathcal{A}_{m-1} - A_{n-1}\mathcal{A}_m)}{(B_{n-1}\mathcal{B}_m - B_n\mathcal{B}_{m-1})t + (A_n\mathcal{B}_{m-1} - A_{n-1}\mathcal{B}_m)}.$$

Since the determinant of this transformation is

$$(A_n B_{n-1} - A_{n-1}B_n)(\mathcal{A}_m\mathcal{B}_{m-1} - \mathcal{A}_{m-1}\mathcal{B}_m) = \pm 1,$$

the proof is complete.

Now we may use (5) to find that

$$t - \frac{A_k}{B_k} = \frac{A_k \zeta_{k+1} + A_{k-1}}{B_k \zeta_{k+1} + B_{k-1}} - \frac{A_k}{B_k} = \frac{-(-1)^{k-1}}{B_k(B_k \zeta_{k+1} + B_{k-1})}$$

and thus the k-th convergent A_k/B_k approximates t with an error of

$$t - \frac{A_k}{B_k} = \frac{(-1)^k}{B_k^2\left(\zeta_{k+1} + \frac{B_{k-1}}{B_k}\right)}. \tag{6}$$

Since $\zeta_{k+1} + B_{k-1}/B_k > 1$ and $B_k = a_k B_{k-1} + B_{k-2} > 2B_{k-2}$ for $k > 2$ implies $B_k > 2^{[k/2]}$, we have that

$$\left| t - \frac{A_k}{B_k} \right| < \frac{1}{2^{2[k/2]}}$$

and thus the convergents approximate t at least exponentially. We shall make more precise statements of this approximation in Chapter IV.

If the partial quotients $a_0, a_1, \ldots, a_k$ are given, we may consider

$$t = [a_0; a_1, a_2, \ldots, a_k, \zeta_{k+1}]$$

as a function of ζ_{k+1}. Formula (6) becomes

$$t(\zeta_{k+1}) = \frac{A_k}{B_k} - \frac{(-1)^{k+1}}{B_k^2\left(\zeta_{k+1} + \dfrac{B_{k-1}}{B_k}\right)}$$

and so t is an increasing function of ζ_{k+1} when $k+1$ is even and decreasing when $k+1$ is odd. We also see that the values of t range between A_k/B_k and $(A_k + A_{k-1})/(B_k + B_{k-1})$.

§3. The transformation $T(x) = \{1/x\}$.

Let $0 < x < 1$ and let the transformation $T:(0,1) \to [0,1)$ be given by $T(x) = \{1/x\}$, the fractional part of $1/x$. Let t be given by (2.2) so that $t = [a_0; a_1, \ldots]$ and we see that $a_0 = [t]$ and $t - a_0 = [0; \zeta_1]$ is in $(0,1)$. Then $T([0; \zeta_1]) = \{\zeta_1\} = [0; \zeta_2]$ and $a_1 = [\zeta_1]$. Similarly, for any $\zeta_k = [a_k; \zeta_{k+1}]$ we have $a_k = [\zeta_k]$ and $T([0; \zeta_k]) = [0; \zeta_{k+1}]$. Let us write $T^k(x)$ for $T(T^{k-1}(x))$ where $k > 1$. By induction, it follows that $T^k([0; \zeta_1]) = [0; \zeta_{k+1}]$ and $a_k = [1/T^{k-1}([0; \zeta_1])]$ for $k > 1$. Of course, if t is rational, both the continued fraction and our process with T^k terminate. Thus we can use the transformation T to recover both the partial and the complete quotients from the number they define.

We also can use T to find a continued fraction representation of any t by setting $a_0 = [t]$ and $x = t - a_0$. If this first remainder is zero, then we are finished while if not, we set $a_1 = [1/x]$. If $T(x) = 0$, we are done while if not, we can set $a_2 = [1/T(x)]$ and so on. If we obtain $T^{n+1}(x) = 0$ at any step, then t is the rational number $[a_0; a_1, \ldots, a_n]$; otherwise, the convergents A_k/B_k converge exponentially to the irrational number t and $a_k = [1/T^{k-1}(\{t\})]$ for $k \geq 1$.

If two continued fractions represent the same number t, must they be identical? Since the initial term must be $[t]$ in both, the first partial quotients must be the same. If they agree up to the k-th partial quotient, then from (2.5) the

$(k+1)$-st complete quotients must be the same (this requires our convention that $a_n \geq 2$ for a terminating continued fraction) so that the $(k+1)$-st partial quotients also must coincide. Thus the regular continued fraction representation of t is unique.

§4. The quantity $D_k = B_k t - A_k$.

Let us define the *k-th difference* D_k by

$$D_k = B_k t - A_k \tag{1}$$

so that $D_{-1} = -1$ and $D_0 = t - a_0 = \{t\}$. Since both the A_k's and B_k's satisfy the recursion formulas (1.2), it immediately follows that

$$D_{k+1} = a_{k+1}D_k + D_{k-1} \quad \text{for } k \geq 0. \tag{2}$$

Further, from our formula (2.6) for $t - A_k/B_k$ we see that

$$D_k = \frac{(-1)^k}{B_k \zeta_{k+1} + B_{k-1}}. \tag{3}$$

Using (2.4) to rewrite ζ_{k+1}, we may transform (3) to find

$$D_k = \frac{(-1)^k}{(a_{k+1}B_k + B_{k-1}) + \dfrac{B_k}{\zeta_{k+2}}} = \zeta_{k+2} \cdot \frac{(-1)^k}{B_{k+1}\zeta_{k+2} + B_k}$$

so that

$$D_{k+1} = -\frac{D_k}{\zeta_{k+2}}. \tag{4}$$

Thus the sequence $\{D_k\}_{k \geq -1}$ of differences alternates in sign and the absolute values of the D_k's approach zero monotonically. From (3) we have

$$|D_k| < \frac{1}{a_{k+1}B_k + B_{k-1}} \leq \tfrac{1}{2} \quad \text{for } k \geq 1, \tag{5}$$

since $B_{k+1} = a_{k+1}B_k + B_{k-1}$ and $B_2 = a_1 a_2 + 1 \geq 2$. Thus $\| B_k t \| = |D_k|$ for

$k \geq 1$. This observation will be important to our discussion of approximation in Chapter II.

§5. Convergents to a number and its reciprocal.

For simplicity, let $t > 1$ be irrational. Then $t = [a_0; a_1, \dots]$ and $0 < 1/t < 1$ is $1/t = [0; a_0, a_1, \dots]$. Let $\{A_k/B_k\}_{k \geq -1}$ be the convergents to t and let $\{\mathcal{A}_k/\mathcal{B}_k\}_{k \geq -1}$ be the convergents to $1/t$. Since

$$A_{-1} = 1 \quad \text{and} \quad B_{-1} = 0$$
$$A_0 = a_0 \quad \text{and} \quad B_0 = 1$$
$$A_1 = a_1 a_0 + 1 \quad \text{and} \quad B_1 = a_1$$

and

$$\mathcal{A}_0 = 0 \quad \text{and} \quad \mathcal{B}_0 = 1$$
$$\mathcal{A}_1 = 1 \quad \text{and} \quad \mathcal{B}_1 = a_0$$
$$\mathcal{A}_2 = a_1 \quad \text{and} \quad \mathcal{B}_2 = a_1 a_0 + 1$$

we have that

$$\mathcal{A}_k = B_{k-1} \quad \text{and} \quad \mathcal{B}_k = A_{k-1}$$

for $k = 0$, 1 and 2. By induction, these formulas hold for all $k \geq 0$ since $\mathcal{A}_{k+1} = a_k \mathcal{A}_k + \mathcal{A}_{k-1} = a_k B_{k-1} + B_{k-2} = B_k$ and similarly for $\mathcal{B}_{k+1}$.

§6. The ratio $\xi_k = B_{k-1}/B_k$.

Since $B_{k+1} = a_{k+1} B_k + B_{k-1}$, the ratio $B_{k+1}/B_k = a_{k+1} + 1/(B_k/B_{k-1})$ and this process may be continued until the ratio $B_1/B_0 = a_1$ is reached. Consequently,

$$\frac{B_{k+1}}{B_k} = [a_{k+1}; a_k, \dots, a_1] \tag{1}$$

for any number t. If $t > 1$, then $a_0 \geq 1$ and the same type of argument may be applied to A_{k+1}/A_k. Of course, if $t < 1$, we can not carry out the process since we would obtain a zero or a negative integer for the last term of the continued fraction and we have excluded such expressions from consideration.

We shall use the notation

$$\xi_k = \frac{B_{k-1}}{B_k}. \tag{2}$$

By (1) we see that

$$\xi_k = [0; a_k, \ldots, a_1]$$

and hence

$$\xi_{k+1} = 1/(a_{k+1} + \xi_k).$$

We may now rewrite (4.3) as

$$D_k = \frac{(-1)^k}{B_k(\zeta_{k+1} + \xi_k)}.$$

§7. The golden ratio and the Fibonacci sequence.

The "golden ratio" is obtained when a line segment or rectangle is divided into two parts such that the ratio between the whole and the larger part is the same as the ratio of this larger part to the remainder. Although Euclid, who referred to this ratio as "division in extreme and mean ratio," was aware that it occurred in many geometric constructions (for example, as the the ratio between the side and base of an isosceles triangle with two 72° base angles), there is no evidence that the ancient Greeks used it as the basis of an artistic or architectural aesthetics. Taking the initial length or area to be 1 and the size of the larger part to be x, and letting g denote the "golden ratio," we have

$$g = \frac{1}{x} = \frac{x}{1-x}.$$

Since $g > 1$, we also have

$$\frac{1}{x} = 1 + \left(\frac{1}{x} - 1\right) = 1 + \frac{1-x}{x} = 1 + \frac{1}{\frac{x}{1-x}}$$

from which it follows that

$$g = 1 + \frac{1}{g} \tag{1}$$

and, consequently, $g = (1 + \sqrt{5})/2$. Repeated application of (1) yields the continued fraction

$$g = 1 + \cfrac{1}{1 + \cfrac{1}{1 + \cdots}}$$

so that $g = [1; 1, 1, 1, \ldots]$.

The sequence $\{B_k\}_{k \geq -1}$ for g is called the *Fibonacci sequence* and usually is denoted by $\{F_k\}_{k \geq 0}$ with $F_{k+1} = B_k$ so that

$$F_0 = 0, \; F_1 = 1, \; F_2 = 1, \; F_3 = 2, \; F_4 = 3, \; F_5 = 5, \; F_6 = 8, \ldots .$$

Since $A_k = B_{k+1}$ for this special number g, we also have that $F_{k+2} = A_k$. The recurrence formulas (1.2) become the famous relation

$$F_{k+1} = F_k + F_{k-1}$$

for the Fibonacci sequence, and the facts that

$$F_{k+2}F_k - F_{k+1}^2 = (-1)^{k+1} \quad \text{and} \quad (F_{k+2}, F_{k+1}) = 1$$

now follow immediately from (1.3) while $F_{k+2}/F_{k+1} \to g$ as $k \to \infty$ by (2.2).

Formula (4.1) becomes $D_k = F_{k+1}g - F_{k+2}$ while (4.4) and the fact that $\zeta_k = g$ for $k \geq 0$ become

$$F_{k+1}g - F_{k+2} = \left(\frac{-1}{g}\right)\left(F_k g - F_{k+1}\right)$$

$$= \cdots = \left(\frac{-1}{g}\right)^k\left(F_1 g - F_2\right) = \left(\frac{-1}{g}\right)^{k+1}(-1)$$

so that $F_{k+1} - gF_k = (-1/g)^k$ and then

$$F_{k+1} = \left(\frac{-1}{g}\right)^k + gF_k. \tag{2}$$

Writing h for $-1/g = (1 - \sqrt{5})/2$, we apply (2) repeatedly to obtain

$$F_{k+1} = h^k + gF_k = h^k + gh^{k-1} + g^2 F_{k-1}$$

$$= \cdots = h^k + gh^{k-1} + g^2 h^{k-2} + \cdots + g^k h^0 + g^{k+1} F_0 = \frac{g^{k+1} - h^{k+1}}{g - h}$$

so that

$$F_{k+1} = \frac{1}{\sqrt{5}} \left(\left(\frac{1+\sqrt{5}}{2} \right)^{k+1} - \left(\frac{1-\sqrt{5}}{2} \right)^{k+1} \right). \tag{3}$$

In Chapter III we shall investigate recurrence relations in more detail and the connection with "periodic" continued fractions will be clarified. The fact that $h = -1/g$ is also the conjugate of g will play an important role.

§8. The continued fraction for e.

The value of e, the base of the natural logarithms, may be estimated from the first few terms of the series representation

$$e = \sum_{k=0}^{\infty} \frac{1}{k!}$$

as $e \doteq 2.718281828$. Expanding this rational approximation as a continued fraction, we find

$$e \doteq [2; 1, 2, 1, 1, 4, 1, 1, 6, 1, 1, 8, 1]$$

and it is then but a small step to conjecture that

$$e = [2; 1, 2, 1, 1, 4, 1, 1, \ldots, 2k, 1, 1, \ldots]; \tag{1}$$

that is, $a_0 = 2$, $a_1 = 1$ and $a_{3k-1} = 2k$, $a_{3k} = 1$ and $a_{3k+1} = 1$ for $k > 0$.

The sequence of convergents thus would begin $2/1$, $3/1$ and then we would have:

k	1	2	3
$\dfrac{A_{3k-1}}{B_{3k-1}}$	$\dfrac{8}{3}$	$\dfrac{87}{32}$	$\dfrac{1264}{465}$
$\dfrac{A_{3k}}{B_{3k}}$	$\dfrac{11}{4}$	$\dfrac{106}{39}$	$\dfrac{1457}{536}$
$\dfrac{A_{3k+1}}{B_{3k+1}}$	$\dfrac{19}{7}$	$\dfrac{193}{71}$	$\dfrac{2721}{1001}$

Since $a_{3k-1} = 2k$ while $a_{3k} = a_{3k+1} = 1$, it is clear that from a purely calculative viewpoint, the convergents A_{3k+1}/B_{3k+1} would give us the most efficient approximations with the least effort. In fact, we can find only these convergents by solving the recursion formulas (1.2) for just these terms:

$$A_{3k+1} = A_{3k} + A_{3k-1} = \left(A_{3k-1} + A_{3k-2}\right) + A_{3k-1}$$

$$= 2A_{3k-1} + A_{3k-2} = 2\left(2kA_{3k-2} + A_{3k-3}\right) + A_{3k-2}$$

$$= \left(2(2k) + 1\right)A_{3k-2} + 2A_{3k-3}$$

$$= \left(2(2k) + 1\right)A_{3k-2} + A_{3k-3} + \left(A_{3k-4} + A_{3k-5}\right)$$

$$= \left(2(2k) + 1\right)A_{3k-2} + \left(A_{3k-3} + A_{3k-4}\right) + A_{3k-5}.$$

Thus

$$A_{3k+1} = 2(2k+1)A_{3(k-1)+1} + A_{3(k-2)+1} \quad \text{for } k \geq 2,$$

and, in a completely similar manner,

$$B_{3k+1} = 2(2k+1)B_{3(k-1)+1} + B_{3(k-2)+1} \quad \text{for } k \geq 2.$$

Let us turn from these numeric speculations to an investigation of some power series arising from combinations of powers of e. Let m be a fixed positive integer. Then

$$e^{1/m} = \sum_{k=0}^{\infty} \frac{1}{k!} \left(\frac{1}{m}\right)^k,$$

$$\frac{1}{2}\left(e^{1/m} + e^{-1/m}\right) = \sum_{k=0}^{\infty} \frac{1}{(2k)!} \left(\frac{1}{m}\right)^{2k},$$

and

$$\frac{1}{2}\left(e^{1/m} - e^{-1/m}\right) = \sum_{k=0}^{\infty} \frac{1}{(2k+1)!} \left(\frac{1}{m}\right)^{2k+1}.$$

Let us consider the positive numbers

$$x_n = \sum_{k=0}^{\infty} \frac{2^n(n+k)!}{k!(2n+2k)!} \left(\frac{1}{m}\right)^{2k+n} \quad \text{for } n \geq 0;$$

x_0 and x_1 are just our last two previous expressions. These numbers satisfy

$$x_n - m(2n+1)x_{n+1} = x_{n+2}$$

since

$$\sum_{k=0}^{\infty} \left(\frac{2^n(n+k)!}{k!(2n+2k)!} \left(\frac{1}{m}\right)^{2k+n} - m(2n+1) \frac{2^{n+1}(n+1+k)!}{k!(2n+2+2k)!} \left(\frac{1}{m}\right)^{2k+n+1} \right)$$

$$= \sum_{k=0}^{\infty} \left(\frac{2^n(n+k)!(2n+1+2k)(2n+2+2k)}{k!(2n+2+2k)!} - \frac{2^{n+1}(2n+1)(n+1+k)}{k!(2n+2+2k)!} \right) \left(\frac{1}{m}\right)^{2k+n}$$

$$= \sum_{k=0}^{\infty} \left(\frac{2^{n+1}(n+1+k)!\big((2n+1+2k)-(2n+1)\big)}{k!(2n+2+2k)!} \right) \left(\frac{1}{m}\right)^{2k+n}$$

$$= \sum_{k=1}^{\infty} \left(\frac{2^{n+2}\big((n+2)+(k-1)\big)!}{(k-1)!\big(2(n+2)+2(k-1)\big)!} \right) \left(\frac{1}{m}\right)^{2(k-1)+(n+2)},$$

and, consequently, the x_n's are monotonically decreasing.

Letting $\zeta_n = x_n/x_{n+1}$, we have that $\zeta_n > 1$,

$$\zeta_0 = \frac{e^{2/m}+1}{e^{2/m}-1}$$

and

$$\zeta_n = \frac{x_{n+2} + m(2n+1)x_{n+1}}{x_{n+1}} = m(2n+1) + \frac{1}{\zeta_{n+1}}.$$

Thus the ζ_n's are the complete quotients of the continued fraction of x_0/x_1 and

$$\frac{e^{2/m}+1}{e^{2/m}-1} = [m;\ 3m,\ 5m,\ \dots,\ (2k+1)m,\ \dots\,].$$

In particular, when $m = 2$ we have that

$$\frac{e+1}{e-1} = [2;\ 6,\ 10,\ 14,\ \dots,\ 2(2k+1),\ \dots\,].$$

Writing $\mathcal{A}_k/\mathcal{B}_k$ for the convergents of $(e+1)/(e-1)$, we see that $\mathcal{A}_0/\mathcal{B}_0 = 2/1$, $\mathcal{A}_1/\mathcal{B}_1 = 13/6$, and

$$\mathcal{A}_{k+1} = 2(2k+1)\mathcal{A}_k + \mathcal{A}_{k-1}$$
$$\mathcal{B}_{k+1} = 2(2k+1)\mathcal{B}_k + \mathcal{B}_{k-1}$$

for $k > 0$. But these recursion formulas are the same as those for every third conjectured convergent of e. Since

$$A_1 = \mathcal{A}_0 + \mathcal{B}_0 \quad \text{and} \quad B_1 = \mathcal{A}_0 - \mathcal{B}_0$$

and
$$A_4 = \mathcal{A}_1 + \mathcal{B}_1 \quad \text{and} \quad B_4 = \mathcal{A}_1 - \mathcal{B}_1,$$

it follows immediately that

$$A_{3k+1} = \mathcal{A}_k + \mathcal{B}_k \quad \text{and} \quad B_{3k+1} = \mathcal{A}_k - \mathcal{B}_k \quad \text{for } k \geq 0.$$

Now we have that

$$\lim_{k \to \infty} \frac{A_k}{B_k} = \lim_{k \to \infty} \frac{A_{3k+1}}{B_{3k+1}} = \lim_{k \to \infty} \frac{\mathcal{A}_k + \mathcal{B}_k}{\mathcal{A}_k - \mathcal{B}_k}$$

$$= \lim_{k \to \infty} \frac{\mathcal{A}_k/\mathcal{B}_k + 1}{\mathcal{A}_k/\mathcal{B}_k - 1} = \frac{\left(\dfrac{e+1}{e-1}\right) + 1}{\left(\dfrac{e+1}{e-1}\right) - 1} = e$$

and the conjecture (1) is proven.

NOTES

Our presentation of the "classical" part of the theory of regular continued fractions is taken from Perron [1954], Dickson [1971], and Khintchine [1963].

§1. The simplest continued fractions (1) with integer partial quotients are the "regular" continued fractions for which a_0 is an arbitrary integer and $a_1, \dots, a_n$ are positive integers. This continued fraction possesses two important properties: no further restrictions on the partial quotients are needed (that is, the expression (1) is well defined for any such choice of the a_k's), and the "best approximations" to the number (1) coincide with the convergents. We devote our attention throughout this book to regular continued fractions. Continued fractions as analytic functions of complex partial quotients are studied in Wall [1973] and in Jones and Thron [1980].

§2 and §3. Regular continued fractions may be developed from several different starting points. We choose to begin with the convergent series (2.2) and then deduce the relation with "best approximations" in the next chapter. It is also possible to begin with the notion of "best approximation" and then construct the continued fraction as the sequence of best approximations. Since this process may be shown to be equivalent to the Euclidean algorithm, regular continued fractions may be defined in terms of the values generated by the steps of the Euclidean algorithm and then the best approximation properties can be deduced. Treatments of regular continued fractions from these other points of view may be found in Cassels [1972], Hardy and Wright [1971], LeVeque [1977], Niven [1956] and Niven and Zuckerman [1980].

The connection between continued fractions and the modular group suggested by Theorems 1 and 2 provides another approach that may be pursued in Rademacher [1983], Chapter 9, and Serre [1973], Chapter VII.

Of course, other types of continued fraction are possible. Since the Euclidean algorithm uses the "greatest integer function" at each iteration (or, the equivalent "fractional part" transformation T described in §3), one natural possibility is to alter the process by applying some other function. Continued fractions "to the nearest integer" are one such possibility and were first studied by Hurwitz in connection with the minima of binary quadratic forms. The convergents of these continued fractions form a subsequence of those of the corresponding regular continued fractions. The metrical theory of these continued fractions may be developed in the same way we shall use in Chapter V (see Rockett [1980]; Rieger [1978] uses the techniques of Lévy to the same end) and estimations of the "relative density" of the nearest integer convergents in the sequence of best approximations then follow (see reference 6 in Rockett [1980]). Regular and nearest integer continued fractions appear as special cases of the "α-continued fractions" studied by Tanaka and Ito [1981].

§7. Markowsky [1992] discusses many erroneous claims about the golden ratio that "... have achieved the status of common knowledge and are widely repeated." Dickson [1971], I, page 393, credits Robert Simson [*Phil. Trans. Roy. Soc. London* **48**, I (1753), 368-376] with the connection between the Fibonacci sequence and the convergents to g.

§8. Formula (1) is due to Euler; our discussion follows Perron [1951], §32.

Chapter II

THE LAW OF BEST APPROXIMATION

§1. Best approximations.

We now are ready to demonstrate the fundamental connection between the theory of regular continued fractions and the study of diophantine approximation (that is, approximations using integers). Perhaps the simplest diophantine approximation problem is the homogeneous problem to minimize the absolute value of the linear expression

$$yt - x \tag{1}$$

where y is a natural number and x is an integer. If t is irrational, then this question is meaningful for both large and small values of y while for rational t, it makes sense only when y is required to be much smaller than the denominator of t. Thus it is natural to require also that y in (1) is bounded. We will call a rational number a/b a *best approximation* of the number t if $|bt - a| < |qt - p|$ for any rational number p/q different from a/b with $0 < q \leq b$. It follows immediately that $(a,b) = 1$ since if $(a,b) = m > 1$, the smaller numbers a/m and b/m would give a better approximation by a factor of m. The definition of best approximation may be rephrased in terms of the distance to the nearest integer because a is the nearest integer to bt and $\|bt\| < \|qt\|$ for any $0 < q \leq b$. This chapter is devoted primarily to two proofs showing that (except for one trivial exception) the best approximations to t coincide with the convergents of t. Although these proofs will depend on estimates of $D_k = B_k t - A_k$, the first will use only that

$$\frac{1}{B_k + B_{k+1}} < |B_k t - A_k| < \frac{1}{B_{k+1}} \tag{2}$$

19

(as we know from (I.4.3) since $a_{k+1} < \zeta_{k+1} < a_{k+1} + 1$) while the second will depend on a more careful analysis of the D_k's introduced in the previous chapter.

§2. The first proof.

We divide the claim into two parts: first, that every best approximation is a convergent and, second, the converse that every convergent is a best approximation. As we shall see, this converse must be refined to remove a class of trivial exceptions.

Theorem 1. *Let t be a real number. Every best approximation to t is a convergent of the regular continued fraction expansion of t.*

Proof. Let $t = [a_0; a_1, a_2, \dots]$. If t is rational, then we require that the final partial quotient $a_n \geq 2$. Since the convergents approximate t alternately from above and below, the situation on the real number line is:

$$a_0 = \frac{A_0}{B_0} \quad \frac{A_2}{B_2} \quad \cdots \quad t \quad \cdots \quad \frac{A_3}{B_3} \quad \frac{A_1}{B_1} \quad a_0 + 1$$

If the best approximation a/b is already a convergent, there is nothing further to be done, so let us suppose that a/b is not a convergent. We shall eliminate all possible places for the number a/b relative to the convergents A_k/B_k.

First, suppose $a/b < A_0/B_0$. Then

$$|bt - a| \;=\; b\left|t - \frac{a}{b}\right| \;\geq\; \left|t - \frac{a}{b}\right| \;>\; |t - a_0|,$$

and this contradicts the assumption that a/b is a best approximation. Second, suppose that $a/b > A_1/B_1$. Since $A_1/B_1 > t$, we have that

$$|bt - a| \;=\; b\left|t - \frac{a}{b}\right| \;>\; b\left|\frac{A_1}{B_1} - \frac{a}{b}\right| \;\geq\; b\,\frac{1}{B_1 b} \;=\; \frac{1}{B_1} \;=\; \frac{1}{a_1}.$$

But $|t - a_0| < 1/a_1$, which again results in a contradiction.

Finally, if a/b lies between A_0/B_0 and A_1/B_1 and is not a convergent, it must lie between two convergents of the same order, say k and $k+2$:

$$\frac{A_0}{B_0} \ \cdots \ \frac{A_{k+1}}{B_{k+1}} \ \cdots \ t \ \cdots \ \frac{A_{k+2}}{B_{k+2}} \ \ \frac{a}{b} \ \ \frac{A_k}{B_k} \ \cdots \ \frac{A_1}{B_1}$$

Thus

$$\frac{1}{B_k B_{k+1}} = \left| \frac{A_{k+1}}{B_{k+1}} - \frac{A_k}{B_k} \right| > \left| \frac{a}{b} - \frac{A_k}{B_k} \right| \geq \frac{1}{B_k b}$$

and so $b > B_{k+1}$. Since

$$|bt - a| = b \left| t - \frac{a}{b} \right| > b \left| \frac{A_{k+2}}{B_{k+2}} - \frac{a}{b} \right| \geq b \frac{1}{B_{k+2} b} = \frac{1}{B_{k+2}},$$

we have lower bounds on both the size of b and the degree of approximation to t by a/b. But $|B_{k+1} t - A_{k+1}| < 1/B_{k+2}$ by (1.2) so a/b could not be a best approximation, and the theorem is established.

We now consider the converse assertion that every convergent is a best approximation. Suppose that the number t is halfway between two integers so that $t = a_0 + 1/2$. Then $|t - (a_0 + 1)| = |t - a_0|$ and our desired converse is false. If we exclude the possibility that $t = a_0 + 1/2$, we may proceed but we shall see that the major portion of the proof is devoted to the investigation of the quantity $|bt - a|$ when bt is halfway between two integers.

Theorem 2. *If t is not of the form $a_0 + 1/2$, then every convergent of t is a best approximation to t.*

Proof. Let B_n be given and let B be that value of b such that $0 < b \leq B_n$ and $|bt - a|$ is minimized (if there are several choices for B, we take the smallest). Let A be the corresponding value for a and let us suppose that A is unique. Then A/B is a best approximation and hence must be a convergent A_k/B_k for some $k \leq n$ by Theorem 1. If $k < n$, we know from (1.2) that

$$|B_k t - A_k| > \frac{1}{B_k + B_{k+1}} \geq \frac{1}{B_{n-1} + B_n},$$

while $|B_n t - A_n| < 1/B_{n+1}$. But if $|B_k t - A_k|$ were less than $|B_n t - A_n|$, we would have $B_{n+1} < B_n + B_{n-1}$, which is impossible. Thus $k = n$ and A/B is the convergent A_n/B_n.

Now we must consider the situation in which A is not unique. Then Bt must be halfway between two integers and we may call them A and $A+1$. Thus $Bt - A = 1/2$ and so $t = (1 + 2A)/(2B)$. Since $(1 + 2A, 2B)$ can not be 2, if $(1 + 2A, 2B) = m > 1$, then $(2B/m)t - (1 + 2A)/m = 0$ would contradict the definition of B, so the rational number $t = (1 + 2A)/(2B)$ has a continued fraction expansion ending as $t = A_n/B_n$ with $A_n = 1 + 2A$ and $B_n = 2B = a_n B_{n-1} + B_{n-2}$, where $a_n \geq 2$. If either $n = 1$ and $a_n > 2$ or $n > 1$, we have $B_{n-1} < B$ so that

$$|B_{n-1}t - A_{n-1}| = \left| B_{n-1}(\frac{A_n}{B_n}) - A_{n-1} \right| = \frac{1}{B_n} = \frac{1}{2B} \leq |Bt - A| = \frac{1}{2},$$

contradicting the definition of B. Since we have excluded the remaining case in which $n = 1$ and $a_n = 2$, our proof is complete.

§3. A theorem of Lagrange.

Since the best approximations to and the convergents of the number t coincide, it is easy to see that a given A_k/B_k must give the smallest possible value of $|bt - a|$ for all choices of a/b for $b > 0$ up to B_{k+1} (since otherwise there would be another best approximation a/b with b between B_k and B_{k+1}). While this follows immediately from Theorems 2.1 and 2.2, it also is possible to begin the study of best approximations with such an observation. Lagrange gave the first proofs from this point of view. We shall give the proof by Legendre of the following theorem of Lagrange.

Theorem 1. *Let a/b be different from A_{k+1}/B_{k+1} with $0 < b \leq B_{k+1}$. Then*

$$|bt - a| \geq |B_k t - A_k| > |B_{k+1}t - A_{k+1}|.$$

Proof. Consider the equation

$$bt - a = m(B_{k+1}t - A_{k+1}) + n(B_k t - A_k).$$

Separating the coefficients of t from the constant terms, we obtain two equations in two unknowns:

$$a = mA_{k+1} + nA_k \quad \text{and} \quad b = mB_{k+1} + nB_k$$

with determinant $A_{k+1}B_k - A_k B_{k+1} = (-1)^k$. Thus m and n must be integers. Since a/b is different from A_{k+1}/B_{k+1}, n can not be zero. If $m = 0$, then

$$|bt - a| = |n(B_k t - A_k)| \geq |B_k t - A_k|.$$

Now suppose both m and n are non-zero. Since $b \leq B_{k+1}$, m and n must be of opposite signs and hence, since $B_{k+1}t - A_{k+1}$ and $B_k t - A_k$ also are of opposite signs,

$$|bt - a| = |m(B_{k+1}t - A_{k+1})| + |n(B_k t - A_k)| \geq |B_k t - A_k|.$$

Finally, we have that $|B_k t - A_k| > |B_{k+1}t - A_{k+1}|$ since we have excluded (see Chapter I) the possibility that $\zeta_{k+1} = 1$.

This proof contains an idea worth exploring further: the approximation capability of a/b was studied by expressing a and b in terms of the A_k's and the B_k's. In the next section, we extend this idea to express b in terms of the B_k's in such a way that the coefficient of each B_k is non-negative and bounded. In the following section, we shall display the power of this method by finding all solutions of the approximation problem $b \| bt \| < K$, where $0 < K \leq 1$ is a given constant.

§4. Ostrowski's algorithm and a second proof.

Although our remarks on best approximation are true for both rational and irrational numbers, for simplicity we now suppose that t is irrational. Thus the sequence of B_k's is $1 = B_0 \leq B_1 < B_2 < B_3 < \cdots$ and for any given natural number

m there is an index N so that $B_N \leq m < B_{N+1}$. We can write m as $m = [m/B_N] \cdot B_N + R$, where the remainder R is $0 \leq R < B_N$. If $R > 0$, this process can be repeated to break R down into a multiple of another B_k, where $k < N$, and another remainder, and then again until the remainder is zero. This decomposition process is *Ostrowski's algorithm* (see Ostrowski [1921]) and it allows us to express m as a sum of multiples of the B_k's for $0 \leq k \leq N$, which we shall call the *Ostrowski representation* of m:

$$m = \sum_{k=0}^{N} c_{k+1} B_k. \tag{1}$$

We notice that the construction of the c_{k+1}'s shows that they are unique and satisfy $0 \leq c_{k+1} \leq a_{k+1}$ for $k > 1$ and $0 \leq c_1 < a_1$ (since $B_0 = 1$ and $B_1 = a_1$). Moreover, $c_k = 0$ if $c_{k+1} = a_{k+1}$, since otherwise we would have had at least $B_{k-1} + a_{k+1} B_k = B_{k+1}$ and c_{k+2} would have been one larger in this downward construction. In the case $t = g$, the representation (1) of m becomes a sum of Fibonacci numbers with the c_{k+1}'s being either 0 or 1 and two consecutive c_{k+1}'s may not both be 1's.

Let us investigate $\| mt \|$. Since the distance to the nearest integer is unchanged by an integer translation,

$$\| mt \| = \left\| mt - \sum_{k=0}^{N} c_{k+1} A_k \right\| = \left\| \sum_{k=0}^{N} c_{k+1} D_k \right\|,$$

where the D_k's are defined by (I.4.1). If this sum of D_k's is small enough, the distance to the nearest integer and the absolute value of our expression will be the same. Since it is much easier to deal with expressions of the latter kind, we may expect this observation to be useful. However, we first need:

Lemma 1. *Let c_{n+1} be the first non-zero c_{k+1} in (1), so that $0 \leq n \leq N$. Then*

$$| (c_{n+1} - 1) D_n - D_{n+1} | < \left| \sum_{k=0}^{N} c_{k+1} D_k \right| < | c_{n+1} D_n - D_{n+1} |.$$

Proof. We know from (I.4.3) that the D_k's alternate in sign so

$$|c_{n+1}D_n + c_{n+2}D_{n+1} + \cdots + c_{N+1}D_N|$$
$$> |c_{n+1}D_n + (a_{n+2}-1)D_{n+1} + a_{n+4}D_{n+3} + \cdots|,$$

since we have omitted all the D_{n+2}, D_{n+4}, ... terms of the same sign as D_n and have included the largest possible multiples of all the terms of the opposite sign. But from (I.4.2), we can write each $a_{k+1}D_k$ as a difference to obtain

$$\left| \sum_{k=0}^{N} c_{k+1}D_k \right| >$$
$$|c_{n+1}D_n - D_{n+1} + (D_{n+2} - D_n) + (D_{n+4} - D_{n+2}) + \cdots|$$
$$= |(c_{n+1}-1)D_n - D_{n+1}|,$$

as claimed. For the upper estimate, we have in a similar way that

$$\left| \sum_{k=0}^{N} c_{k+1}D_k \right| < |c_{n+1}D_n + a_{n+3}D_{n+2} + a_{n+5}D_{n+4} + \cdots|,$$

and the result follows.

We are now in position to state our main result connecting the distance to the nearest integer with the absolute value of the sum of D_k's.

Theorem 1. *Let t be an irrational number and $m > 1$ be a positive integer. Let the Ostrowski representation of m given by (1) have $c_{k+1} = 0$ for $0 \le k < n \le N$ and $c_{n+1} > 0$. Then*

(1) *if $n \ge 2$, then $\| mt \| = \left| \sum_{k=0}^{N} c_{k+1}D_k \right|$;*

(2) *if $\{t\} < 1/2$, then (a) $\| mt \| = \left| \sum_{k=0}^{N} c_{k+1}D_k \right|$ if $c_1 = 0$ and $c_2 > 0$;*

and (b) $\| mt \| > |D_1|$ if $c_1 > 0$; and

(3) *if $\{t\} > 1/2$, then $c_1 = 0$ for any m and (a) $\| mt \| > \| t \|$ if $c_2 > 1$; and (b) $\| mt \| > D_2$ if $c_2 = 1$.*

Proof. Let us write S for $|\sum c_{k+1} D_k|$. Then for part (1) we have from Lemma 1 that

$$S < |c_{n+1} D_n - D_{n+1}| \le |a_{n+1} D_n - D_{n+1}| = |-D_{n-1}| \le |-D_1| < \tfrac{1}{2},$$

by (I.4.5), so $\|mt\| = S$. For part (2a),

$$S < |c_2 D_1 - D_2| \le |a_2 D_1 - D_2| = |-D_0| = t - a_0 < \tfrac{1}{2}$$

and again $\|mt\| = S$. For part (2b), we see that

$$|-D_1| < |(c_1 - 1) D_0 - D_1| < S$$

$$< |c_1 D_0 - D_1| \le |(a_1 - 1) D_0 - D_1| = |-D_{-1} - D_0|,$$

because $c_1 < a_1$. Since $D_0 = \{t\}$ we have $1 - a_1 \{t\} < \|mt\| < 1 - \{t\}$ and our claim follows. For part (3), we have $a_1 = 1$ so that $c_1 = 0$. In part (3a), we have

$$|D_1| < |D_1 - D_2| < S < |c_2 D_1 - D_2| \le |a_2 D_1 - D_2| = |-D_0|,$$

so $1 - \{t\} < S < \{t\}$ and the result follows. Finally, for part (3b), we have $|-D_2| < S < |D_1 - D_2|$ from the Lemma. Since $|-D_2| = D_2$ and $|D_1 - D_2| = D_2 - D_1$, our result will follow if $1 - (D_2 - D_1) > D_2$, which is the same as $1 + D_1 > 2 D_2$. But $t = [a_0; 1, \zeta_2]$, by (I.2.3), and $\zeta_2 < 2a_2$ so that $t < [a_0; 1, 2a_2]$ and hence $2a_2 > (2a_2 + 1)(t - a_0)$. Since $D_0 = t - a_0$ and $D_1 = (t - a_0) - 1$, we obtain $1 - 2D_0 > (2a_2 - 1) D_1$ and thus $1 + D_1 > 2a_2 D_1 + 2D_0 = 2D_2$, making our proof complete.

With this result, we may proceed with a second proof of the fact that the best approximations to t and the convergents of t coincide.

Theorem 2. *Let t be irrational. Then A/B is a best approximation to t if and only if it is a convergent of t.*

Proof. We apply Theorem 1 with $m = B$ and examine $\| Bt \|$. Suppose first that $n \geq 2$ in the Ostrowski representation of B. Then $\| Bt \|$ will be minimized exactly when $n = N$ and $c_{N+1} = 1$, and thus $B = B_N$ and $A = A_N$. Our result follows similarly when B and t satisfy condition (2a) of Theorem 1. It remains to consider the special cases arising from (2b), (3a) and (3b) of Theorem 1.

In (2b), we have $\{t\} < 1/2$ and $c_1 > 0$. If $c_1 = 1$ and $N = 0$, then $B = 1$, $\| Bt \| = \| t \| = D_0$, and so $B = B_0$ and $A = A_0$. Otherwise, we have $B > 1$ and $\| Bt \| > | - D_1 |$, and such a B can be neither a best approximation nor a convergent. In (3a), it is clear that such a B can be neither a best approximation nor a convergent. Finally, in (3b) we have $\{t\} > 1/2$ and $c_1 = 0$ for any B. If $c_2 = 1$ and $N = 1$, then $B = 1$, $\| Bt \| = \| t \| = | - D_1 |$, and hence $B = B_1$ and $A = A_1$. Otherwise, we have $B > 1$ and $\| Bt \| > D_2$, and such a B can be neither a best approximation nor a convergent, and our proof is finished.

§5. The approximation $b \| bt \| < K$.

Since $1/B_{k+1} < 1/B_k$, we have from (1.2) that $\| B_k t \| < 1/B_k$. We now study all solutions of this kind of relation. Let $0 < K \leq 1$ and consider $b \| bt \| < K$; that is,

$$| bt - a | < \frac{K}{b}.$$

Suppose that $(a, b) = 1$. When will a/b be a best approximation to t? Suppose that $0 < q \leq b$ and that $| qt - p | < | bt - a |$. Then $| t - p/q | < K/(bq)$ and $| t - a/b | < K/b^2$ combine to give

$$\frac{1}{bq} \leq \left| \frac{p}{q} - \frac{a}{b} \right| \leq \left| t - \frac{p}{q} \right| + \left| t - \frac{a}{b} \right| < \frac{K}{b} \left(\frac{1}{q} + \frac{1}{b} \right)$$

and so

$$b \left(\frac{1}{K} - 1 \right) < q.$$

If $1/2 < K \leq 1$, this is not impossible, but for $0 < K \leq 1/2$ we have $b < q$, contrary to the assumption that $q \leq b$. In this case, we have that a/b is a convergent of t since it is a best approximation. For $0 < K \leq 1$, if $(a, b) = c > 1$, we have $a = cp$ and $b = cq$ with $(p, q) = 1$ and then $| bt - a | < K/b$ implies that $| qt - p |$

$< K/(c^2 q)$, which means that p/q must be a convergent of t and hence a and b are of the form $a = cA_k$ and $b = cB_k$ for some index k. For such numbers, $b \parallel bt \parallel < K$ is the same as

$$\frac{c}{B_k(\zeta_{k+1} + \xi_k)} < \frac{K}{cB_k},$$

using the notation of (I.6.2), and so $c^2 < K(\zeta_{k+1} + \xi_k)$. As all solutions of $b \parallel bt \parallel < K$ are of this form for $0 < K \leq 1/2$, we have shown:

Theorem 1. *Let* $0 < K \leq 1/2$. *If* $b \parallel bt \parallel < K$, *then* $b = cB_k$, *where* $c > 0$ *satisfies* $c^2 < K(\zeta_{k+1} + \xi_k)$.

Of course, we have not established that there are any such solutions since it may be that $K(\zeta_{k+1} + \xi_k) < 1$. We shall return to this question in Chapter IV and again in Chapter VI. We now turn to the case $K = 1$. The solution for any K with $1/2 < K < 1$ may be found by a similar analysis.

Theorem 2. *If* $b \parallel bt \parallel < 1$, *then* b *must satisfy either*

    (1) $b = c_{k+1} B_k$, *where* $0 < c_{k+1} < \sqrt{\zeta_{k+1} + \xi_k}$, *or*

    (2) $b = B_k + c_{k+2} B_{k+1}$, *where* $c_{k+2} = 1$ *if* $\zeta_{k+2} < 2 + B_{k+1}/B_k$ *or* $c_{k+2} = a_{k+2} - 1$ *if* $\zeta_{k+3} + 2 > B_{k+2}/B_{k+1}$.

Proof. Let b have an Ostrowski representation as in (4.1). By Lemma 4.1, we have at least that $|D_{n+1}| < 1/b$ and thus $b < B_{n+1}\zeta_{n+2} + B_n$. Since $B_N \leq b < B_{N+1}$, $n \leq N$ and we can not have $n \leq N-3$, because then b would be less than $B_{N-3} + B_{N-2}\zeta_{N-1}$, which is less than B_N. Thus any solution of $b \parallel bt \parallel < 1$ must have $n \geq N-2$, and so must be of the form

$$b = c_{k+1} B_k + c_{k+2} B_{k+1} + c_{k+3} B_{k+2}.$$

As in the demonstration of Theorem 1, if $b = cB_k$ then $c^2 < \zeta_{k+1} + \xi_k$. Since $\zeta_{k+1} > 1$, $c = 1$ is always possible and part (1) is established.

Suppose that $b = c_{k+1} B_k + c_{k+2} B_{k+1}$, where c_{k+1} and $c_{k+2} > 0$. Then

$$b \parallel bt \parallel \; = \; \left(c_{k+1}B_k + c_{k+2}B_{k+1}\right)|D_k|\left(c_{k+1} - \frac{c_{k+2}}{\zeta_{k+2}}\right)$$

$$= \; \frac{c_{k+2}B_{k+1} + c_{k+1}B_k}{B_{k+1} + B_k/\zeta_{k+2}}\left(c_{k+1} - \frac{c_{k+2}}{\zeta_{k+2}}\right).$$

If $c_{k+1} > 1$, both terms are greater than 1 and hence $b \parallel bt \parallel \; > 1$. It remains to consider the case $c_{k+1} = 1$:

$$b \parallel bt \parallel \; = \; \frac{B_{k+1} + B_k/c_{k+2}}{B_{k+1} + B_k/\zeta_{k+2}}\, c_{k+2}\left(1 - \frac{c_{k+2}}{\zeta_{k+2}}\right).$$

Since $c_{k+2} < \zeta_{k+2}$, the first fraction is greater than 1 and we must have that $c_{k+2}(1 - c_{k+2}/\zeta_{k+2}) < 1$. If $c_{k+2} \leq a_{k+2} - 2$, then $c_{k+2} < \zeta_{k+2} - 2$,

$$c_{k+2}\left(1 - \frac{c_{k+2}}{\zeta_{k+2}}\right) \; < \; c_{k+2}\left(\frac{2}{\zeta_{k+2}}\right),$$

and thus $c_{k+2} < \zeta_{k+2}/2$. But then

$$c_{k+2}\left(1 - \frac{c_{k+2}}{\zeta_{k+2}}\right) \; < \; \frac{c_{k+2}}{2}$$

and so $c_{k+2} < 2$. Thus we need consider only $c_{k+2} = 1$ or $a_{k+2} - 1$. If $c_{k+2} = 1$, we must have

$$\left(\zeta_{k+2} - 1\right)\left(B_k + B_{k+1}\right) \; < \; B_k + \zeta_{k+2}B_{k+1}$$

so that

$$\zeta_{k+2} \; < \; 2 + \frac{B_{k+1}}{B_k},$$

while if $c_{k+2} = a_{k+2} - 1$, we see that $B_k + c_{k+2}B_{k+1} = B_{k+2} - B_{k+1}$ and we must have

$$\left(\zeta_{k+2} - a_{k+2} + 1\right)\left(B_{k+2} - B_{k+1}\right) \; < \; B_{k+2} + \frac{B_{k+1}}{\zeta_{k+3}},$$

which means that

$$\frac{B_{k+2}}{B_{k+1}} - 2 \; < \; \zeta_{k+3},$$

since $\zeta_{k+2} - a_{k+2} = 1/\zeta_{k+3}$.

We conclude the proof by showing that b can not be of the form

$$b = c_{k+1}B_k + c_{k+2}B_{k+1} + c_{k+3}B_{k+2},$$

where both c_{k+1} and c_{k+3} are positive. If $c_{k+2} = 0$, then

$$
\begin{aligned}
b\,\|\,bt\,\| &= \left(c_{k+3}B_{k+2} + c_{k+1}B_k\right)|\,D_k\,|\left(c_{k+1} + \frac{c_{k+3}}{\zeta_{k+2}\zeta_{k+3}}\right) \\
&= \frac{c_{k+3}B_{k+2} + c_{k+1}B_k}{B_{k+1} + B_k/\zeta_{k+2}}\left(c_{k+1} + \frac{c_{k+3}}{\zeta_{k+2}\zeta_{k+3}}\right),
\end{aligned}
$$

and this is greater than 1. If $c_{k+2} > 0$, then

$$
\begin{aligned}
b\,\|\,bt\,\| &= b\,|\,D_k\,|\left(c_{k+1} - \frac{c_{k+2}}{\zeta_{k+2}} + \frac{c_{k+3}}{\zeta_{k+2}\zeta_{k+3}}\right) \\
&> \frac{b}{B_{k+1} + B_k/\zeta_{k+2}}\left(\frac{c_{k+1}\zeta_{k+2} - c_{k+2}}{\zeta_{k+2}}\right) = \frac{b(c_{k+1}\zeta_{k+2} - c_{k+2})}{\zeta_{k+2}B_{k+1} + B_k},
\end{aligned}
$$

and this is greater than 1 since $a_{k+2} + 1 > \zeta_{k+2}$.

It should be noted that the conditions on ζ_{k+2} and ζ_{k+3} in Theorem 2 are not mutually exclusive and numbers can be constructed having one, both or none of these properties. In particular, from the continued fraction (I.8.1) for e we may show:

Corollary. *If $b\,\|\,be\,\| < 1$, then b is either 2 or cB_k, where $0 < c < \sqrt{\zeta_{k+1} + \xi_k}$.*

Proof. We need to consider only condition (2) of Theorem 2. Since $c_{k+2} < a_{k+2}$, we can exclude those k's for which $a_{k+2} = 1$. If $a_{k+2} > 1$, the requirement that $\zeta_{k+2} < 2 + a_{k+1} + \xi_k < 4$ forces $k+2 = 2$ so then $c_2 = 1$ and $b = B_0 + c_2 B_1 = 2$, while the requirement that $\xi_{k+3} + 2 > a_{k+2} + \xi_{k+1}$ contradicts $\zeta_{k+3} < 2$, since $a_{k+3} = 1$.

The previous methods may be applied to solve any approximation problem $b\,\|\,bt\,\| < K$ where $K > 0$. It is clear that as K increases, the length of the expressions to be considered also will increase and that expressions of a given length will give approximations within an error that depends only on the length. Clearly, $\| bt \|$ is minimized for integers b of the form

$$B_n + (a_{n+2}-1)B_{n+1} + a_{n+4}B_{n+3} + \cdots + a_{n+2k}B_{n+2k-1} = B_{n+2k} - B_{n+1},$$

and we have for such numbers that

$$b \| bt \| > b| - D_{n+1}| > \frac{B_{n+2k} - B_{n+1}}{B_{n+2} + B_{n+1}} = \frac{\frac{B_{n+2k}}{B_{n+2}} - \xi_{n+2}}{1 + \xi_{n+2}}.$$

But

$$\frac{B_{n+2k}}{B_{n+2}} = \frac{B_{n+2k}}{B_{n+2k-1}} \cdot \frac{B_{n+2k-1}}{B_{n+2k-2}} \cdot \ldots \cdot \frac{B_{n+3}}{B_{n+2}} > g^{2k-3},$$

since the sequence $\{B_k\}_{k \geq 0}$ increases slowest for $t = g$, and thus

$$b \| bt \| > \frac{g^{2k-3} - 1}{2}.$$

If $b \| bt \| < K$, we must have

$$\frac{g^{2k-3} - 1}{2} < K$$

and we have proven:

Theorem 3. *Let $K > 0$ and let t be an irrational number. Then all positive integers b such that $b \| bt \| < K$ have the form*

$$c_{n+1}B_n + c_{n+2}B_{n+1} + \cdots + c_{n+m}B_{n+m-1}$$

with length m such that $m < (\log_g(2K + 1) + 3)/2$, and the coefficients c_{k+1} satisfy $0 \leq c_1 < a_1$, $0 \leq c_{k+1} \leq a_{k+1}$ for $k > 1$, and $c_k = 0$ if $c_{k+1} = a_{k+1}$.

A lower bound on the length m would require an upper bound on the ratios B_{k+1}/B_k and these numbers may be unbounded (as, for example, is the case with the number e). We shall see in Chapter III that an upper bound can be found if t is a quadratic surd and we will give the resulting extension of Theorem 3 in our discussion of Pell's equation in Chapter IV.

§6. The *t*-expansion of a real number.

We have seen that the Ostrowski representation of a natural number immediately yields information about the minimum of the linear homogeneous form (1.1), since this representation shows how the number can be broken into parts, each being a best approximation to the number t. In this section, we shall develop an analogous tool to study the corresponding inhomogeneous problem of minimizing the absolute value of

$$yt - x + s,\tag{1}$$

where s is an arbitrary real number. Of course, if s is of the form $s = Bt - A$ for integers A and B, then (1) reduces to the homogeneous problem. For an s not of this form, it is natural to seek a representation showing how s can be broken into parts, each being the degree of approximation achieved by a best approximation to t. This amounts to repeating the Ostrowski representation using the differences $\{D_k\}_{k \geq 0}$ in place of the denominators $\{B_k\}_{k \geq 0}$ of the convergents.

However, there is a sign problem which must be addressed before we proceed. In the homogeneous case (1.1), we could neglect negative values for the integer y, since $\| -yt \| = \| yt \|$, but it is not true that $\| -yt + s \| = \| yt + s \|$ for an arbitrary s. We shall see in Chapter IV that the minimum of $|y| \, \| yt + s \|$ depends on whether y represents an arbitrary integer or a positive integer.

The Ostrowski representation (4.1) is for a natural number m. For a negative integer y, there certainly exists an index N as before such that $-B_{N+1} \leq y < -B_N$ and hence $y + B_{N+1}$ is a non-negative integer m. If $m > 0$, then m has the representation (4.1) and we may write y as

$$y = \Big(\sum_{k=0}^{N} c_{k+1} B_k \Big) - B_{N+1},\tag{2}$$

which we also will call the "Ostrowski representation" of the (negative) integer y. Since $B_{N+3} = a_{N+3} B_{N+2} + B_{N+1}$, we also have that

$$y = \Big(\sum_{k=0}^{N} c_{k+1} B_k \Big) + a_{N+3} B_{N+2} - B_{N+3},$$

$$y = \Big(\sum_{k=0}^{N} c_{k+1}B_k \Big) + a_{N+3}B_{N+2} + a_{N+5}B_{N+4} - B_{N+5},$$

and so on. With these alternative expressions for (2) in mind, let us now consider "Ostrowski sums" of D_k's instead of B_k's.

Theorem 1. *Let t be an irrational number and let a_k, ζ_k, A_k, B_k and D_k refer to the continued fraction of t. Any real number s not of the form $s = Bt - A$ for integers A and B and such that*

$$\frac{-1}{\zeta_1} < s < 1 - \frac{1}{\zeta_1}$$

has a unique t-expansion

$$s = \sum_{k=0}^{\infty} c_{k+1}D_k, \tag{3}$$

where $0 \le c_1 < a_1$, $0 \le c_{k+1} \le a_{k+1}$ for $k > 1$, and $c_k = 0$ if $c_{k+1} = a_{k+1}$.

Proof. We first note that if $c_{k+1} = 0$ for all sufficiently large k's, then

$$s = \Big(\sum_{k=0}^{N} c_{k+1}B_k \Big) t - \Big(\sum_{k=0}^{N} c_{k+1}A_k \Big) = Bt - A$$

where $B > 0$. On the other hand, if $s = Bt - A$ where $B < 0$, then by (2) we have that

$$s = \sum_{k=0}^{N} c_{k+1}D_k - D_{N+1}$$

$$= \sum_{k=0}^{N} c_{k+1}D_k + a_{N+3}D_{N+2} + a_{N+5}D_{N+4} + \cdots , \tag{4}$$

and these last expressions will be excluded during our construction of the t-expansion (3) of s. We also note that the requirement that s be in the interval $(-1/\zeta_1, 1 - 1/\zeta_1)$ of length 1 is sufficient for applications to (1), because any real number may be rewritten as an integer translate of an s in our interval.

Since $-1/\zeta_1 = -D_0$ and $1 - 1/\zeta_1 = (a_1 - 1)D_0 - D_1$, we begin with a real number s satisfying

$$-D_0 < s < (a_1 - 1)D_0 - D_1;$$

that is, using the same method as in the proof of Lemma 4.1,

$$a_2 D_1 + a_4 D_3 + \cdots < s < (a_1 - 1)D_0 + a_3 D_2 + a_5 D_4 + \cdots$$

and, by our initial remarks, the endpoints have been excluded from further consideration by the requirements on s. Since $0 \leq c_1 < a_1$, we must consider expressions of the form

$$s = c_1 D_0 + \left(\sum_{k=1}^{\infty} c_{k+1} D_k \right) = c_1 D_0 + s'$$

where either $c_1 = 0$, in which case

$$a_2 D_1 + a_4 D_3 + \cdots \leq s \leq a_3 D_2 + a_5 D_4 + \cdots$$

and so, excluding the endpoints by our previous remarks,

$$-D_0 < s < -D_1; \tag{5a}$$

or $0 < c_1 < a_1$, in which case

$$c_1 D_0 + (a_2 - 1)D_1 + a_4 D_3 + \cdots \leq s \leq c_1 D_0 + a_3 D_2 + a_5 D_4 + \cdots$$

and thus, again excluding the endpoints for the same reason,

$$(c_1 - 1)D_0 - D_1 < s < c_1 D_0 - D_1. \tag{5b}$$

The system of open intervals (5a and 5b) partitions the original interval for s into $a_1 - 1$ open sub-intervals, each separated from the next by an excluded number of the form (4) and each corresponding to a unique value of c_1.

Now suppose $s = c_1 D_0 + s'$ is given. Can we repeat the above process to determine c_2? If $c_1 = 0$, we must show that the possible values $0 \leq c_2 \leq a_2$ fill out the interval $(-D_0, -D_1)$ in the same manner as before. If $c_2 = 0$, then

$$0D_0 + 0D_1 + a_4 D_3 + a_6 D_5 + \cdots < s < 0D_0 + 0D_1 + a_3 D_2 + a_5 D_4 + \cdots$$

so
$$-D_2 < s < -D_1,$$

while if $0 < c_2 \leq a_2$,

$$0D_0 + c_2D_1 + a_4D_3 + a_6D_5 + \cdots < s < 0D_0 + c_2D_1 + (a_3 - 1)D_2 + a_5D_4 + \cdots$$

so
$$c_2D_1 - D_2 < s < (c_2 - 1)D_1 - D_2.$$

Since $c_2 = a_2$ gives a lower endpoint of $a_2D_1 - D_2 = a_2D_1 - (a_2D_1 + D_0) = -D_0$, we obtain a partition of $(-D_0, -D_1)$ with the same structure as previously, and thus c_2 is uniquely determined in the case that $c_1 = 0$. If $c_1 > 0$, we must show that the possible values $0 \leq c_2 < a_2$ fill out the interval $((c_1 - 1)D_0 - D_1, c_1D_0 - D_1)$ in the same manner as above. If $c_2 = 0$, then

$$c_1D_0 + 0D_1 + a_4D_3 + a_6D_5 + \cdots < s < c_1D_0 + 0D_1 + a_3D_2 + a_5D_4 + \cdots$$

so
$$c_1D_0 - D_2 < s < c_1D_0 - D_1,$$

while if $0 < c_2 < a_2$,

$$c_1D_0 + c_2D_1 + a_4D_3 + a_6D_5 + \cdots < s < c_1D_0 + c_2D_1 + (a_3 - 1)D_2 + a_5D_4 + \cdots$$

so
$$c_1D_0 + c_2D_1 - D_2 < s < c_1D_0 + (c_2 - 1)D_1 - D_2.$$

Since $c_2 = a_2 - 1$ gives a lower endpoint of $c_1D_0 + (a_2 - 1)D_1 - D_2 = (c_1 - 1)D_0 - D_1$, we again obtain a similar partition of the starting interval, and c_2 has been uniquely determined in the case that $c_1 > 0$.

The argument for c_{k+1} given the coefficients $c_1, \ldots, c_k$ proceeds precisely as for the case of c_2 given c_1 detailed above, except that the parity of k determines whether the sub-intervals fill out the starting interval from left to right or from right to left.

After we develop some other needed materials, we shall apply these t-expansions in Chapter IV to the inhomogeneous problem (1).

NOTES

§1. Our definition of a best approximation is sometimes termed a "best approximation of the second kind" with a "best approximation of the first kind" being a fraction a/b (where $b > 0$) such that $|t - a/b| < |t - p/q|$ for any fraction $p/q \neq a/b$ with $0 < q \leq b$. Unlike Theorems 2.1 and 2.2, it is difficult to characterize the best approximations of the first kind. Clearly, a best approximation of the second kind is also of the first kind. It can be shown (see, for instance, Khintchine [1963]) that every best approximation of the first kind is either a convergent or of the form $(A_k + c_{k+2}A_{k+1})/(B_k + c_{k+2}B_{k+1})$, provided that the index $k = -1$ is now allowed; unfortunately, the converse statement does not hold (for example, consider 5/2 and 7/3 as approximations to 29/12).

§2. This is the "usual" proof; see, for instance, §6 of Khintchine [1963].

§3. See Perron [1954], §15.

§4. Although the representation (1) was proven and used by Ostrowski [1921] in his study of diophantine approximation (see also Lang [1966a,b]), it was rediscovered by authors (obviously unaware of Ostrowski's paper) a generation later for the special case $t = g$ (see Lekkerkerker [1952]) and then another generation later (see Fraenkel [1985], where other references may be found).

§5. Theorem 1 is due to Legendre (see Perron [1954], §13). The numbers $A_k + c_{k+2}A_{k+1}$ and $B_k + c_{k+2}B_{k+1}$ arising in part (2) of Theorem 2 are relatively prime since, in general,

$$(A_k + (c+1)A_{k+1})(B_k + cB_{k+1}) - (A_k + cA_{k+1})(B_k + (c+1)B_{k+1}) = (-1)^k$$

for $c = 0, 1, \ldots, a_{k+2} - 1$. The quotients $(A_k + c_{k+2}A_{k+1})/(B_k + c_{k+2}B_{k+1})$ are sometimes called "quasi-convergents" (see LeVeque [1977]), "intermediate convergents" (see Khintchine [1963]) or "auxiliary convergents" (see Grace [1918a]) and the A_k/B_k are then termed "principal convergents." There does not seem to be any terminology for the longer expressions such as $B_k + c_{k+2}B_{k+1} + c_{k+3}B_{k+2}$ considered in the proof of Theorem 2 or in the statement of Theorem 3. Theorem 2 is due to Grace [1918a]. The Corollary is equivalent to Adams [1966], in which he shows that the only reduced fractions a/b satisfying $|e - a/b| < 1/b^2$ are the convergents except for a finite number of exceptions. Theorem 3 is from Rockett and Szüsz [1986].

§6. Theorem 1 is equivalent to the representation given as Theorem 1 in Sós [1958], which, in turn, is essentially the algorithm used by Davenport [1947] and by Descombes [1956a,b,c] in their studies of the inhomogeneous problem (1). We believe that our formulation seems a natural outgrowth of the Ostrowski representation of §4, rather than an ad hoc method developed for a particular instance of (1).

Chapter III

PERIODIC CONTINUED FRACTIONS

§1. The classical theorems.

In §7 of Chapter I, we saw that the "golden ratio" g satisfied the equation

$$g \ = \ 1 + \frac{1}{g}$$

and thus was the continued fraction $[1; 1, 1, \dots]$ as well as a root of the quadratic equation $x^2 - x - 1 = 0$. In general, a *quadratic surd* is a solution of a quadratic equation

$$ax^2 + bx + c \ = \ 0 \tag{1}$$

with integer coefficients a, b and c where a is not zero and the discriminant $D = b^2 - 4ac$ is not a perfect square. Since (1) may be rewritten as a relation between x and its reciprocal, we may expect that quadratic surds and continued fractions showing a repeating characteristic should be related.

A *purely periodic* continued fraction with *period* (or length) m is a continued fraction such that the initial block of partial quotients $a_0, \dots, a_{m-1}$ is repeated over and over (so that $a_m = a_0, \dots, a_{2m-1} = a_{m-1}$, and so on for each a_{km} where $k \geq 1$) and no shorter block $a_0, \dots, a_{n-1}$ with $n < m$ has this property. We will write a purely periodic continued fraction as

$$[\overline{a_0; a_1, \dots, a_{m-1}}]. \tag{2}$$

A *periodic* continued fraction consists of an *initial block* of length n followed by a *repeating block* of length m and will be written as

39

$$[a_0; a_1, \ldots, a_{n-1}, \overline{a_n, \ldots, a_{n+m-1}}]. \tag{3}$$

Again, we suppose that there is no shorter such repeating block and that the initial block does not end with a copy of the repeating block.

We begin with a theorem of Euler.

Theorem 1. *If t is a periodic continued fraction, then t is a quadratic surd.*

Proof. Suppose first that t is purely periodic and has the form given by (2). Then $t = \zeta_0 = \zeta_m = \zeta_{2m} = \ldots$ and by (I.2.5) we have

$$t = \frac{A_{m-1}t + A_{m-2}}{B_{m-1}t + B_{m-2}}$$

and so

$$B_{m-1}t^2 + (B_{m-2} - A_{m-1})t - A_{m-2} = 0. \tag{4}$$

Since $m > 0$, $B_{m-1} \geq B_0 = 1$ and thus t is a quadratic surd. Suppose now that t has initial terms and is of the form (3). Then $\zeta_n = \zeta_{n+m} = \zeta_{n+2m} = \ldots$ and we can use (I.2.5) twice to obtain

$$t = \frac{A_{n-1}\zeta_n + A_{n-2}}{B_{n-1}\zeta_n + B_{n-2}} \quad \text{and} \quad t = \frac{A_{n+m-1}\zeta_n + A_{n+m-2}}{B_{n+m-1}\zeta_n + B_{n+m-2}}$$

and then solve for ζ_n:

$$\zeta_n = -\frac{B_{n-2}t - A_{n-2}}{B_{n-1}t - A_{n-1}} \quad \text{and} \quad \zeta_n = -\frac{B_{n+m-2}t - A_{n+m-2}}{B_{n+m-1}t - A_{n+m-1}}.$$

After cross multiplying and collecting terms, we find that

$$(B_{n-2}B_{n+m-1} - B_{n-1}B_{n+m-2})t^2$$
$$+ (-B_{n-2}A_{n+m-1} - A_{n-2}B_{n+m-1} + A_{n-1}B_{n+m-2} + B_{n-1}A_{n+m-2})t$$
$$+ (A_{n-2}A_{n+m-1} - A_{n-1}A_{n+m-2}) = 0.$$

If this were not a quadratic equation, then $B_{n-2}B_{n+m-1}$ would equal $B_{n-1}B_{n+m-2}$ and then B_{n+m-1} would divide $B_{n-1}B_{n+m-2}$. Since B_{n+m-1} and

B_{n+m-2} are relatively prime, B_{n+m-1} would have to divide B_{n-1}, and this is impossible since $B_{n+m-1} > B_{n-1}$. Thus t is a quadratic surd.

The converse of Theorem 1 was first shown by Lagrange although the result was known earlier. We give two proofs: the first, due to Charves, for its brevity, and the second, due to Lagrange, for its method. The two step proof of Charves first observes that the unimodular relation between t and ζ_{k+1} given by (I.2.5) means that if t satisfies a quadratic equation, ζ_{k+1} satisfies a quadratic equation with the same discriminant, and then uses the approximation of t by A_k/B_k to deduce that the quadratic equations for $t = \zeta_0$, ζ_1, ζ_2, ... have bounded coefficients and thus must eventually repeat. While based on the same principles, Lagrange's proof investigates both the ζ_k's and their conjugates, and this material will be developed further in our subsequent study of purely periodic continued fractions.

Theorem 2. *If t is a quadratic surd, then t is a periodic continued fraction.*

First proof. Since t is a quadratic surd, there are integers $p_0 > 0$, q_0 and r_0 such that $p_0 t^2 + q_0 t + r_0 = 0$. By using (I.2.5) to replace t with an expression in an arbitrary ζ_{k+1}, we find

$$p_0(A_k\zeta_{k+1} + A_{k-1})^2 + q_0(A_k\zeta_{k+1} + A_{k-1})(B_k\zeta_{k+1} + B_{k-1})$$
$$+ r_0(B_k\zeta_{k+1} + B_{k-1})^2 = 0.$$

After expanding and collecting terms, this becomes

$$p_{k+1}\zeta_{k+1}^2 + q_{k+1}\zeta_{k+1} + r_{k+1} = 0, \tag{5}$$

where

$$p_{k+1} = p_0 A_k^2 + q_0 A_k B_k + r_0 B_k^2$$
$$q_{k+1} = 2p_0 A_k A_{k-1} + q_0(A_k B_{k-1} + A_{k-1}B_k) + 2r_0 B_k B_{k-1}$$
$$r_{k+1} = p_0 A_{k-1}^2 + q_0 A_{k-1}B_{k-1} + r_0 B_{k-1}^2$$

so that $r_{k+1} = p_k$. If p_{k+1} were 0, ζ_{k+1} would be a rational number and then so would t, hence p_{k+1} is not 0 for all $k \geq 0$. A straight forward calculation shows that the discriminant is

$$q_{k+1}^2 - 4p_{k+1}r_{k+1} = (q_0^2 - 4p_0r_0)(A_kB_{k-1} - A_{k-1}B_k)^2$$

and thus remains unchanged.

Since $|B_kt - A_k| < 1/B_{k+1} < 1/B_k$, we may write $A_k = B_kt + \epsilon/B_k$ where $|\epsilon| < 1$. Then the formula for p_{k+1} becomes

$$p_{k+1} = p_0\left(B_kt + \frac{\epsilon}{B_k}\right)^2 + q_0\left(B_kt + \frac{\epsilon}{B_k}\right)B_k + r_0B_k^2$$

$$= (p_0t^2 + q_0t + r_0)B_k^2 + \epsilon(2p_0t + q_0) + p_0\left(\frac{\epsilon}{B_k}\right)^2 = \epsilon(2p_0t + q_0) + p_0\left(\frac{\epsilon}{B_k}\right)^2$$

and we see that

$$|p_{k+1}| < |2p_0t| + |q_0| + |p_0|,$$

which means that $|p_{k+1}|$ is bounded. Since $r_{k+1} = p_k$, $|r_{k+1}|$ also is bounded as then is q_{k+1}, since the discriminant is constant. Thus the coefficients of all the equations (5) are bounded and there can be only finitely many such equations. Consequently, the equations will have to repeat and the corresponding ζ_{k+1}'s will coincide. Thus t is a periodic continued fraction as was to be shown.

Second proof. By the quadratic formula, either solution of (1) can be written as

$$t = \frac{P_0 + \sqrt{D}}{Q_0}$$

where $Q_0 \neq 0$ and P_0 are integers, the integer $D > 0$ is not a perfect square and Q_0 divides $D - P_0^2$. To develop t as a continued fraction, we set $a_0 = [t]$ and then

$$\zeta_1 = \frac{1}{t - a_0} = \frac{Q_0}{(P_0 - a_0Q_0) + \sqrt{D}} = \frac{P_1 + \sqrt{D}}{Q_1},$$

where

$$P_1 = a_0Q_0 - P_0$$

and

$$Q_1 = \frac{D - P_0^2 + 2a_0 P_0 Q_0 - (a_0 Q_0)^2}{Q_0}$$

are integers. Since $(D - P_1^2)/Q_1 = Q_0$, P_1 and Q_1 possess the same divisibility property noticed for P_0 and Q_0. Continuing this process, we find

$$\zeta_{k+1} = \frac{P_{k+1} + \sqrt{D}}{Q_{k+1}} \tag{6}$$

where $P_{k+1} = a_k Q_k - P_k$ and $Q_{k+1} = (D - P_{k+1}^2)/Q_k$ is not zero since D is not a perfect square.

Let η_k be the *conjugate* of ζ_k; that is,

$$\eta_k = \bar{\zeta}_k = \frac{P_k - \sqrt{D}}{Q_k}.$$

By conjugating formula (I.2.5), we find

$$\eta_0 = \frac{A_k \eta_{k+1} + A_{k-1}}{B_k \eta_{k+1} + B_{k-1}}$$

and thus

$$\eta_{k+1} = -\frac{B_{k-1}\eta_0 - A_{k-1}}{B_k \eta_0 - A_k} = -\frac{B_{k-1}}{B_k} \cdot \frac{\eta_0 - (A_{k-1}/B_{k-1})}{\eta_0 - (A_k/B_k)}.$$

Since $A_k/B_k \to t$ as $k \to \infty$, we see that $\eta_{k+1} = -\xi_k(1 + \epsilon_k)$, where $\epsilon_k \to 0$ as $k \to \infty$. It follows that for k sufficiently large, $-1 < \eta_{k+1} < 0$. Since $\zeta_{k+1} > 1$, we can make the following sequence of observations for large k's:

$$\zeta_k - \eta_k > 0 \quad \text{so} \quad \frac{2\sqrt{D}}{Q_k} > 0 \quad \text{so} \quad Q_k > 0$$

$$\zeta_k + \eta_k > 0 \quad \text{so} \quad \frac{2P_k}{Q_k} > 0 \quad \text{so} \quad P_k > 0$$

$$\eta_k < 0 \quad \text{so} \quad P_k - \sqrt{D} < 0 \quad \text{so} \quad P_k < \sqrt{D}$$

$$\eta_k > -1 \quad \text{so} \quad \frac{P_k - \sqrt{D}}{Q_k} > -1 \quad \text{so} \quad \sqrt{D} - P_k < Q_k$$

$$\zeta_k > 1 \qquad \text{so} \qquad \frac{P_k + \sqrt{D}}{Q_k} > 1 \qquad \text{so} \qquad Q_k < \sqrt{D} + P_k.$$

Thus P_k and Q_k are positive integers with $0 < P_k < \sqrt{D}$ and $0 < \sqrt{D} - P_k < Q_k < \sqrt{D} + P_k$. Since there are only a finite number of possible values for the pairs P_k and Q_k, the values must repeat and thus the continued fraction is periodic from some point on.

We also have that $(P_k + \sqrt{D})/Q_k > \sqrt{D}$ implies $P_k > \sqrt{D}(Q_k - 1)$ while $P_k < \sqrt{D}$. Thus $\zeta_k > \sqrt{D}$ if and only if $Q_k = 1$ and we see that the partial quotients a_k are $< \sqrt{D}$ unless $Q_k = 1$, in which case $a_k < 2\sqrt{D}$.

For example, if t is the quadratic surd $(7 + \sqrt{11})/19$, we may take $Q_0 = 19$, $P_0 = 7$ and $D = 11$, since 19 is a divisor of $11 - 7^2$. Thus $\zeta_0 = (7 + \sqrt{11})/19$ and $a_0 = [\zeta_0] = 0$. Solving for

$$P_{k+1} = a_k Q_k - P_k, \quad Q_{k+1} = \frac{D - P_{k+1}^2}{Q_k},$$

$$\zeta_{k+1} = \frac{P_{k+1} + \sqrt{D}}{Q_{k+1}} \quad \text{and} \quad a_{k+1} = [\zeta_{k+1}]$$

in order, we have:

$$P_0 = 7, \quad Q_0 = 19, \quad \zeta_0 = \frac{7 + \sqrt{11}}{19}, \quad a_0 = 0$$

$$P_1 = 0 \cdot 19 - 7 = \text{-}7, \quad Q_1 = \frac{11 - (\text{-}7)^2}{19} = \text{-}2, \quad \zeta_1 = \frac{\text{-}7 + \sqrt{11}}{\text{-}2}, \quad a_1 = 1$$

$$P_2 = 1 \cdot (\text{-}2) - (\text{-}7) = 5, \quad Q_2 = \frac{11 - 5^2}{\text{-}2} = 7, \quad \zeta_2 = \frac{5 + \sqrt{11}}{7}, \quad a_2 = 1$$

$$P_3 = 1 \cdot 7 - 5 = 2, \quad Q_3 = \frac{11 - 2^2}{7} = 1, \quad \zeta_3 = \frac{2 + \sqrt{11}}{1}, \quad a_3 = 5$$

$$P_4 = 5 \cdot 1 - 2 = 3, \quad Q_4 = \frac{11 - 3^2}{1} = 2, \quad \zeta_4 = \frac{3 + \sqrt{11}}{2}, \quad a_4 = 3.$$

Since $\zeta_4 > 1$ and $-1 < \eta_4 < 0$, "k sufficiently large" for this example means "$k > 3$" **and** a_4 starts the periodic block. Continuing,

$$P_5 = 3 \cdot 2 - 3 = 3, \quad Q_5 = \frac{11 - 3^2}{2} = 1, \quad \zeta_5 = \frac{3 + \sqrt{11}}{1}, \quad a_5 = 6$$

$$P_6 = 6 \cdot 1 - 3 = 3, \quad Q_6 = \frac{11 - 3^2}{1} = 2, \quad \zeta_6 = \frac{3 + \sqrt{11}}{2}, \quad a_6 = 3,$$

so $\zeta_6 = \zeta_4$ and the periodic block is $\overline{a_4, a_5}$. Thus

$$\frac{7 + \sqrt{11}}{19} = [0; 1, 1, 5, \overline{3, 6}]$$

as expected. When $k = 5$, $Q_k = 1$ and then "$a_k < 2\sqrt{d}$" becomes $6 < 2\sqrt{11}$, so that a_5 is as large as possible in this example.

It should be clear that the essential idea of the second proof for Theorem 2 was the estimate that $-1 < \eta_k < 0$ for k sufficiently large. We shall call a quadratic surd ζ_0 *reduced* if $\zeta_0 > 1$ and its conjugate η_0 satisfies $-1 < \eta_0 < 0$. The next two theorems are due to Galois.

Theorem 3. *Let t be a quadratic surd. Then t is purely periodic if and only if t is reduced.*

Proof. Suppose that t is purely periodic and has the form (2). Since the value of a_0 appears again as a_m, $a_0 \geq 1$ and so $t = \zeta_0 > 1$. Let

$$f(x) = B_{m-1}x^2 + (B_{m-2} - A_{m-1})x - A_{m-2}.$$

By (4) in the proof of Theorem 1, $f(t) = 0$. Since $f(0) = -A_{m-2} < 0$ while

$$f(-1) = (B_{m-1} - B_{m-2}) + (A_{m-1} - A_{m-2}) > 0,$$

$f(x)$ must have another zero between -1 and 0. But this other root is η_0 and t is reduced.

Now suppose that t is reduced; that is, $\zeta_0 > 1$ and $-1 < \eta_0 < 0$. Since $\eta_0 = a_0 + 1/\eta_1$, we have

$$\frac{1}{\eta_1} = -a_0 + \eta_0 < -a_0 \leq -1$$

and thus $-1 < \eta_1 < 0$. It follows by induction that $-1 < \eta_k < 0$ for $k \geq 0$. By way of contradiction, let us suppose that t is not purely periodic and has the form (3) with $n \geq 1$. Then a_{n-1} is not equal to a_{n+m-1}, because otherwise the period would have begun one position sooner. However, $\zeta_n = \zeta_{n+m}$ and so

$$\zeta_{n-1} - \zeta_{n+m-1} = (a_{n-1} + \frac{1}{\zeta_n}) - (a_{n+m-1} + \frac{1}{\zeta_{n+m}}) = a_{n-1} - a_{n+m-1}$$

is a non-zero integer. Similarly, $\eta_{n-1} - \eta_{n+m-1}$ is a non-zero integer, yet $\eta_{n-1} - \eta_{n+m-1}$ is between -1 and 1. This contradiction establishes the result.

Theorem 4. *Let* $\zeta_0 = [\overline{a_0; a_1, \ldots, a_{m-1}}]$. *Then* $-1/\eta_0$ *has the inverse period form* $-1/\eta_0 = [\overline{a_{m-1}; a_{m-2}, \ldots, a_0}]$.

Proof. We have

$$\zeta_0 = a_0 + \frac{1}{\zeta_1}, \ \zeta_1 = a_1 + \frac{1}{\zeta_2}, \ldots, \zeta_k = a_k + \frac{1}{\zeta_{k+1}}, \ldots, \zeta_{m-1} = a_{m-1} + \frac{1}{\zeta_0}$$

so conjugation gives

$$\eta_0 = a_0 + \frac{1}{\eta_1}, \ \eta_1 = a_1 + \frac{1}{\eta_2}, \ldots, \eta_k = a_k + \frac{1}{\eta_{k+1}}, \ldots, \eta_{m-1} = a_{m-1} + \frac{1}{\eta_0}.$$

Solving these equations in reverse order, we have

$$\frac{-1}{\eta_0} = a_{m-1} - \eta_{m-1}, \ldots, \frac{-1}{\eta_{k+1}} = a_k - \eta_k, \ldots, \frac{-1}{\eta_1} = a_0 - \eta_0.$$

Thus each $-1/\eta_k$ is greater than 1 and we may think of them as the complete quotients of the number $-1/\eta_0 = [\overline{a_{m-1}; a_{m-2}, \ldots, a_0}]$.

We use these last two theorems to prove a result due to Legendre.

Theorem 5. *Let $d > 1$ be a rational number that is not the square of another rational number. Then*

$$\sqrt{d} = [a_0; \overline{a_1, a_2, \ldots, a_2, a_1, 2a_0}]. \tag{7}$$

Proof. Since $\sqrt{d}$ is a quadratic surd, it must be of the form (3). But then $a_0 = [\sqrt{d}]$ and $1/\zeta_1 = \sqrt{d} - [\sqrt{d}]$. It follows that the conjugate of $\zeta_1 > 1$ is $\eta_1 = -1/(\sqrt{d} + [\sqrt{d}])$ and satisfies $-1 < \eta_1 < 0$. Thus ζ_1 is reduced and therefore is purely periodic by Theorem 3. Let us write ζ_1 as $\zeta_1 = [\overline{a_1; a_2, \ldots, a_{m-1}}]$ and so, by Theorem 4, $-1/\eta_1 = [\overline{a_{m-1}; a_{m-2}, \ldots, a_1}]$. Since $\sqrt{d} = \zeta_0 = [a_0; \zeta_1]$, we have two expressions for $\sqrt{d} + [\sqrt{d}]$:

$$\sqrt{d} + [\sqrt{d}] = \zeta_0 + a_0 = [2a_0; \overline{a_1, a_2, \ldots, a_{m-1}}]$$

and

$$\sqrt{d} + [\sqrt{d}] = \frac{-1}{\eta_1} = [a_{m-1}; \overline{a_{m-2}, \ldots, a_1, a_{m-1}}].$$

Thus $a_{m-1} = 2a_0$, $a_1 = a_{m-2}, \ldots, a_k = a_{m-k-1}, \ldots, a_{m-2} = a_1$ and the continued fraction is of the claimed form.

For example, to find the continued fraction for $\sqrt{11/7}$, we rewrite it as

$$\sqrt{\frac{11}{7}} = \frac{0 + \sqrt{77}}{7}$$

since then $Q_0 = 7$, $P_0 = 0$ and $D = 77$ satisfy $Q_0 \mid (D - P_0^2)$. Proceeding as in the previous example, we obtain

$$\sqrt{\frac{11}{7}} = [1; \overline{3, 1, 16, 1, 3, 2}]$$

where

k	0	1	2	3	4	5	6
P_k	0	7	5	8	8	5	7
Q_k	7	4	13	1	13	4	7

and the period length is even. Similarly, we also have

$$\sqrt{\frac{29}{17}} = [1; \overline{3, 3, 1, 2, 1, 14, 14, 1, 2, 1, 3, 3, 2}]$$

where

k	0	1	2	3	4	5	6	7	8	9	10	11	12	13
P_k	0	17	19	14	13	11	20	22	20	11	13	14	19	17
Q_k	17	12	11	27	12	31	3	3	31	12	27	11	12	17

and the period length is odd. The symmetry properties of the P_k's and Q_k's suggested by these two examples may be established by the same method used above for the partial quotients.

We saw in Theorem 3 that if $-1 < \eta_0 < 0$, ζ_0 is purely periodic while the proof of Theorem 5 shows that if $\eta_0 < -1$, then ζ_0 has one initial term before the repeating block. It also may be shown that if $\eta_0 > 0$, there are one or more terms in the initial term before the repeating block.

We close this section with a refinement of Theorem 5 that will be important for a subsequent application in Chapter IV.

Theorem 6. *Let $D > 1$ be an integer that is not a perfect square and let m be the length of the periodic block in the continued fraction (7) for $\sqrt{D}$. Then with the notation of (6), $Q_k = 1$ if and only if $m \mid k$.*

Proof. Since there are m terms, $\zeta_1 = \zeta_{m+1}$ while $\zeta_0, \zeta_1, \ldots, \zeta_m$ are distinct. Thus

$$\zeta_m = 2a_0 + \frac{1}{\zeta_{m+1}} = 2a_0 + \frac{1}{\zeta_1} = 2a_0 + (\sqrt{D} - a_0) = a_0 + \sqrt{D}.$$

But $\zeta_m = (P_m + \sqrt{D})/Q_m$ and if Q_m were greater than 1, then the integer $P_m - a_0 Q_m$ would equal the irrational number $(Q_m - 1)\sqrt{D}$. Thus $Q_m = 1$. A similar argument shows that $Q_k = 1$ for every multiple of m.

Now suppose that $Q_k = 1$. Then $\zeta_k = P_k + \sqrt{D}$. Since ζ_k is purely periodic, $-1 < \eta_k < 0$ and so $1 > \sqrt{D} - P_k > 0$. Thus $P_k < \sqrt{D} < P_k + 1$ and $P_k = [\sqrt{D}] = a_0$. Hence $\zeta_k = \zeta_m$, and k must be a multiple of m.

In the same way as the previous numerical illustrations, we have (taking $Q_0 = 1$ and $P_0 = 0$):

$$\sqrt{29} = [5; \overline{2, 1, 1, 2, 10}]$$

with

k	0	1	2	3	4	5
P_k	0	5	3	2	3	5
Q_k	1	4	5	5	4	1

$$\sqrt{34} = [5; \overline{1, 4, 1, 10}]$$

with

k	0	1	2	3	4
P_k	0	5	4	4	5
Q_k	1	9	2	9	1

$$\sqrt{77} = [8; \overline{1, 3, 2, 3, 1, 16}]$$

with

k	0	1	2	3	4	5	6
P_k	0	8	5	7	7	5	8
Q_k	1	13	4	7	4	13	1

and

$$\sqrt{493} = [22; \overline{4, 1, 10, 3, 3, 10, 1, 4, 44}]$$

with

k	0	1	2	3	4	5	6	7	8	9
P_k	0	22	14	19	21	18	21	19	14	22
Q_k	1	9	33	4	13	13	4	33	9	1

Although the periodic block for $\sqrt{77}$ is the same except for the starting position as that of $\sqrt{11/7}$ (which has the same discriminant of 77), this is not an indication of a general rule as seen by comparing $\sqrt{493}$ and $\sqrt{29/17}$. We shall refer to these examples again in Chapter IV.

§2. Period lengths.

We have seen that any quadratic surd is a periodic continued fraction, possibly with an initial block, and it is natural to ask how the length of the repeating block depends on the surd. In the second proof of Theorem 1.2, we expressed the surd in the form

$$t = \frac{P_0 + \sqrt{D}}{Q_0}$$

and found that

$$\zeta_{k+1} = \frac{P_{k+1} + \sqrt{D}}{Q_{k+1}}$$

where P_{k+1} and Q_{k+1} are positive integers satisfying

$$D - P_{k+1}^2 \;=\; Q_{k+1}Q_k \qquad\qquad (1)$$

and, for sufficiently large k's, such that

$$0 \;<\; P_{k+1} \;<\; \sqrt{D}$$

and

$$0 \;<\; \sqrt{D} - P_{k+1} \;<\; Q_{k+1} \;<\; \sqrt{D} + P_{k+1} \;<\; 2\sqrt{D}.$$

Since the inequalities $0 < P_{k+1} < \sqrt{D}$ and $0 < Q_{k+1} < 2\sqrt{D}$ insure that there can be no more than $\sqrt{D} \cdot 2\sqrt{D}$ distinct pairs $\{P_{k+1}, Q_{k+1}\}$, we immediately obtain Lagrange's estimate that the period length L of any surd of discriminant D satisfies

$$L(D) \;<\; 2D.$$

We now make a more precise statement about L by combining the inequalities with the relation (1).

Theorem 1. *Let $t = (P_0 + \sqrt{D})/Q_0$ be any quadratic surd where $Q_0 \neq 0$, P_0 and $D > 0$ are integers such that D is not a perfect square and $Q_0 \,|\, (D - P_0^2)$. Then the length L of the repeating block in the periodic continued fraction of t satisfies*

$$L(D) \;=\; \mathcal{O}\!\left(\sqrt{D}\,\log(D)\right). \qquad\qquad (2)$$

Proof. From our previous discussion, it suffices to estimate the number of pairs $\{P, Q\}$ of integers satisfying

$$0 \;<\; P \;<\; \sqrt{D},$$
$$0 \;<\; \sqrt{D} - P \;<\; Q \;<\; \sqrt{D} + P \;<\; 2\sqrt{D}$$

and

$$P^2 \;\equiv\; D \bmod Q.$$

If $Q > \sqrt{D}$, then $\sqrt{D} < Q < \sqrt{D} + P < 2\sqrt{D}$ and hence

$$0 \;<\; Q - \sqrt{D} \;<\; P \;<\; \sqrt{D},$$

while if $Q < \sqrt{D}$, $0 < \sqrt{D} - P < Q < \sqrt{D}$ gives

$$0 < \sqrt{D} - Q < P < \sqrt{D}.$$

In either case, given a value of Q, the possible values for P are contained in an interval of length less than Q, and we can count the number of possible P's by counting their residue classes modulo Q. Thus we have that

$$L(D) < \sum_{0 < Q < 2\sqrt{D}} \left(\sum_{P^2 \equiv D \bmod Q} 1 \right).$$

We now investigate the inner sum

$$\sum_{P^2 \equiv D \bmod Q} 1$$

for the three possible cases $(Q,D) = 1$, $1 < (Q,D) < Q$ and $(Q,D) = Q$. In the first case, let us suppose that $P^2 \equiv D \bmod Q$ is solvable, that is, D is a *quadratic residue* of Q, and let the prime factorization of Q be

$$Q = 2^{k_0} \cdot p_1^{k_1} \cdot \cdots \cdot p_m^{k_m}$$

where $p_1, \ldots, p_m$ are distinct odd primes with positive integer exponents $k_1, \ldots, k_m$ and the integer exponent of 2 is $k_0 \geq 0$. Then any solution of $P^2 \equiv D \bmod Q$ must also satisfy the congruences

$$P^2 \equiv D \bmod 2^{k_0}, \; P^2 \equiv D \bmod p_1^{k_1}, \ldots, P^2 \equiv D \bmod p_m^{k_m}.$$

If we know how many solutions N_i there are to each of these congruences ($i = 0, 1, \ldots, m$) then, by the "Chinese remainder theorem," there are exactly $N_0 \cdot N_1 \cdot \cdots \cdot N_m$ solutions to the original congruence.

For any of the odd primes p_i, a solution of $P^2 \equiv D \bmod p_i^{k_i}$ means there is a solution of $P^2 \equiv D \bmod p_i$, and there are exactly two such solutions. By "Euler's lemma," these solutions "pull back" to two solutions modulo $p_i^{k_i}$ and so $N_i = 2$ for $i = 1, \ldots, m$. For the prime 2, the situation is slightly different: an odd number a

is always a square modulo 2, a is a square modulo 4 only if $a \equiv 1 \mod 4$, and a is a square modulo 2^{k_0} for $k_0 \geq 3$ only if $a \equiv 1 \mod 8$. Thus, given a solution of $P^2 \equiv D \mod 2$, there will be two solutions modulo 4 and four solutions modulo 2^{k_0} for $k_0 \geq 3$. For the case $(Q,D) = 1$, we now have that

$$\sum_{P^2 \equiv D \bmod Q} 1 \; = \; 2^k \cdot 2^m,$$

where $k = 0$ if $k_0 \leq 1$, $k = 1$ if $k_0 = 2$ and $k = 2$ if $k_0 \geq 3$, and m is the number of distinct odd prime divisors of Q. Since $2^{k+m} \leq \tau(Q)$, where $\tau(n)$ is number of divisors of the integer n, we have shown that

$$\sum_{P^2 \equiv D \bmod Q} 1 \; \leq \; \tau(Q).$$

The contribution to our sum for a Q in the second case, when $1 < (Q,D) < Q$, can be no more than that from the first case since for these Q's the congruence $P^2 \equiv D \mod Q$ is the same as

$$(Q,D)\left(\frac{P}{(Q,D)}\right)^2 \; \equiv \; \frac{D}{(Q,D)} \; \bmod \; \frac{Q}{(Q,D)},$$

which either has no solutions or at most $\tau(Q/(Q,D))$ solutions, as we saw in the first case. Finally, if $Q \mid D$, then

$$\sum_{P^2 \equiv D \bmod Q} 1 \; = \; \mathcal{O}(\sqrt{Q}).$$

Since

$$\sum_{k=1}^{n} \tau(k) \; = \; \sum_{d=1}^{n}\left[\frac{n}{d}\right] \; = \; n \log(n) + \mathcal{O}(n),$$

our estimate that

$$L \; = \; \mathcal{O}\!\left(\sum_{0 < Q < 2\sqrt{D}} \tau(Q)\right)$$

is equivalent to (2).

§3. Second order linear recurrences.

A *linear recurrence* $\{x_k\}_{k \geq 0}$ consists of two initial values, x_0 and x_1, and recurrence formulas

$$x_{k+1} = a_{k-1}x_k + x_{k-1} \tag{1}$$

defining x_{k+1} for $k \geq 1$ in terms of the previous two values, where the coefficients a_{k-1} are positive integers. We first show that the elements of any linear recurrence may be expressed as a linear combination of the numerators and denominators of the continued fraction determined by the coefficients $\{a_k\}_{k \geq 0}$.

Lemma 1. *Let $\{x_k\}_{k \geq 0}$ be the linear recurrence (1) and let $\{A_k/B_k\}_{k \geq -1}$ be the convergents of $t = [a_0; a_1, a_2, \ldots]$. Then*

$$x_{k+1} = x_1 A_{k-1} + x_0 B_{k-1} \tag{2}$$

for $k \geq 0$.

Proof. When $k = 0$, this is just $x_1 = x_1 \cdot 1 + x_0 \cdot 0$, while when $k = 1$, $x_2 = a_0 x_1 + x_0$ by (1), which is the same as (2). Given that (2) has been established up to $k + 1$, $x_{k+2} = a_k x_{k+1} + x_k$ becomes $a_k(x_1 A_{k-1} + x_0 B_{k-1}) + (x_1 A_{k-2} + x_0 B_{k-2})$ $= x_1(a_k A_{k-1} + A_{k-2}) + x_0(a_k B_{k-1} + B_{k-2})$ and the result follows by (I.1.2).

For example, suppose $x_0 = 7$, $x_1 = 13$ and the a_{k-1}'s in (1) are the partial quotients for $e = [2; 1, 2, 1, 1, 4, 1, 1, 6, 1, 1, 8, 1, 1, \ldots]$. Then

$$x_{12} = x_1 A_{10} + x_0 B_{10}$$

by (2) and, from §8 of Chapter I, we immediately have

$$x_{12} = 13 \cdot 2721 + 7 \cdot 1001 = 42380,$$

as may be checked by application of the recurrence formulas (1).

We remark that (2) is the same as

$$\frac{x_{k+1}}{B_{k-1}} = x_1 \frac{A_{k-1}}{B_{k-1}} + x_0$$

so that

$$\lim_{k\to\infty} \frac{x_{k+1}}{B_{k-1}} = x_1 t + x_0$$

and thus the asymptotic behavior of the linear recurrence depends only on the initial values and t. Consequently, it suffices to study only the sequence $\{B_k\}_{k \geq -1}$ of denominators. In Chapter V we shall obtain an estimate for the behavior of the B_k's for "most" of the numbers $t \in [0,1)$, but we now examine the B_k's for a class of periodic continued fractions to obtain the general form of formula (I.7.3) for the Fibonacci sequence. As in that particular case, we shall use formula (I.4.4) to decompose D_k into a product of ζ_k's, and so we first investigate such products for a periodic continued fraction.

Lemma 2. *Let* $t = [\overline{a_0; a_1, \ldots, a_{m-1}}]$ *be a purely periodic continued fraction with period* m. *Then*

$$\zeta_0 \cdot \zeta_1 \cdot \cdots \cdot \zeta_{m-1} = B_{m-1}t + B_{m-2}. \tag{3}$$

Of course, the companion formula

$$\eta_0 \cdot \eta_1 \cdot \cdots \cdot \eta_{m-1} = B_{m-1}\bar{t} + B_{m-2}$$

follows immediately by conjugation.

Proof. Since $t = \zeta_0 = \zeta_m$, we have

$$\zeta_0 \cdot \zeta_1 \cdot \cdots \cdot \zeta_{m-1} = \zeta_1 \cdot \zeta_2 \cdot \cdots \cdot \zeta_{m-1} \cdot \zeta_m = \left(B_0\zeta_1 + B_{-1}\right) \cdot \zeta_2 \cdot \cdots \cdot \zeta_{m-1} \cdot \zeta_m$$

$$= \left(\left(B_0(a_1 + \frac{1}{\zeta_2}) + B_{-1}\right) \cdot \zeta_2\right) \cdot \zeta_3 \cdot \cdots \cdot \zeta_{m-1} \cdot \zeta_m = \left(B_1\zeta_2 + B_0\right) \cdot \zeta_3 \cdot \cdots \cdot \zeta_{m-1} \cdot \zeta_m$$

$$= \cdots = \left(B_{m-2}\zeta_{m-1} + B_{m-3}\right) \cdot \zeta_m = B_{m-1}t + B_{m-2},$$

as required.

In our analysis of the Fibonacci sequence, we also made use of the fact that $A_k = B_{k+1}$ for $g = [\overline{1}]$, since this meant that facts about D_k yielded information about $B_k g - B_{k+1}$. We make our generalization by requiring that the periodic continued fraction possesses a similar property.

Theorem 1. *Let* $t = [\overline{a_0;\, a_1,\, \ldots,\, a_{m-1}}]$ *be a purely periodic continued fraction with a symmetric period, so that the sequence of numbers* $a_0, a_1, \ldots, a_{m-1}$ *is the same as* $a_{m-1}, a_{m-2}, \ldots, a_0$, *and let* $\alpha = B_{m-1} t + B_{m-2}$. *Then*

$$B_{km-1} = B_{m-1}\left(\frac{\alpha^k - \overline{\alpha}^k}{\alpha - \overline{\alpha}}\right) \tag{4}$$

for $k \geq 1$.

Since the $m - 1$ denominators between any two consecutive terms given by (4) can easily be recovered from the values of B_{km-1}, $B_{(k+1)m-1}$ and the numbers $a_0, a_1, \ldots, a_{m-1}$, this result suffices to determine all the denominators for the number t. For $t = g$ (so that $m = 1$), formula (4) immediately reduces to that given in §7 of Chapter I for the Fibonacci sequence.

Proof. Let $\zeta_0 = [\overline{a_0;\, a_1,\, \ldots,\, a_{m-1}}]$ and $\eta_0 = \overline{\zeta}_0$ as usual. By (1.4),

$$B_{m-1}\zeta_0^2 + (B_{m-2} - A_{m-1})\zeta_0 - A_{m-2} = 0$$

so that

$$\zeta_0 = \frac{(A_{m-1} - B_{m-2}) + \sqrt{(A_{m-1} - B_{m-2})^2 + 4A_{m-2}B_{m-1}}}{2B_{m-1}}.$$

Since the periodic block is symmetric, $-1/\eta_0 = \zeta_0$ by Theorem 1.4 and thus

$$\frac{(A_{m-1} - B_{m-2}) + \sqrt{(A_{m-1} - B_{m-2})^2 + 4A_{m-2}B_{m-1}}}{2B_{m-1}}$$

$$= \frac{-2B_{m-1}}{(A_{m-1} - B_{m-2}) + \sqrt{(A_{m-1} - B_{m-2})^2 + 4A_{m-2}B_{m-1}}}$$

which means that

$$(A_{m-1} - B_{m-2})^2 - \left((A_{m-1} - B_{m-2})^2 + 4A_{m-2}B_{m-1}\right) = -4B_{m-1}^2,$$

and so $A_{m-2} = B_{m-1}$.

Formula (I.4.1) becomes $D_{km-2} = B_{km-2}t - B_{km-1}$, while (I.4.4) allows us to write

$$B_{km-2}t - B_{km-1} = \left(\frac{-1}{\zeta_{km-1}}\right)\left(B_{km-3}t - A_{km-3}\right)$$

$$= \cdots = \left(\frac{-1}{\zeta_{km-1}}\right)\left(\frac{-1}{\zeta_{km-2}}\right)\cdots\left(\frac{-1}{\zeta_{(k-1)m}}\right)\left(B_{(k-1)m-2}t - B_{(k-1)m-1}\right)$$

which, by (3), is

$$= \overline{\alpha}\left(B_{(k-1)m-2}t - B_{(k-1)m-1}\right) = \cdots = \overline{\alpha}^{k-1}\left(B_{m-2}t - B_{m-1}\right)$$

and so

$$B_{km-2} = \tfrac{1}{t}\left(B_{km-1} + \overline{\alpha}^{k-1}(B_{m-2}t - B_{m-1})\right). \tag{5}$$

From (I.1.2) and the symmetry of the periodic block, we also have that

$$B_{km-1} = a_{km-1}B_{km-2} + B_{km-3} = A_0 B_{km-2} + A_{-1}B_{km-3}$$

$$= A_0\left(a_{km-2}B_{km-3} + B_{km-4}\right) + A_{-1}B_{km-3} = A_1 B_{km-3} + A_0 B_{km-4}$$

$$= \cdots = A_{m-1}B_{(k-1)m-1} + A_{m-2}B_{(k-1)m-2}.$$

Using (5) for $B_{(k-1)m-2}$ and substituting into the above expression, we have that

$$B_{km-1} = A_{m-1}B_{(k-1)m-1} + A_{m-2}\left(\tfrac{1}{t}\left(B_{km-1} + \overline{\alpha}^{k-2}\left(B_{m-2}t - B_{m-1}\right)\right)\right)$$

so that

$$B_{km-1} = \left(A_{m-1} + \tfrac{1}{t}A_{m-2}\right)B_{(k-1)m-1} + \left(\tfrac{1}{t}A_{m-2}\left(B_{m-2}t - B_{m-1}\right)\right)\overline{\alpha}^{k-2}. \tag{6}$$

But

$$A_{m-1} + \tfrac{1}{t}A_{m-2} = \tfrac{1}{t}\left(A_{m-1}t + A_{m-2}\right)$$

$$= \tfrac{1}{t}\left(\frac{A_{m-1}t + A_{m-2}}{B_{m-1}t + B_{m-2}}\right)\left(B_{m-1}t + B_{m-2}\right) = \alpha$$

and

$$\tfrac{1}{t}A_{m-2}\big(B_{m-2}t - B_{m-1}\big) = B_{m-1}\big(B_{m-2} - \tfrac{1}{t}B_{m-1}\big)$$

$$= B_{m-1}\big(B_{m-1}\eta_0 + B_{m-2}\big) = B_{m-1}\overline{\alpha},$$

so that (6) is really

$$B_{km-1} = \alpha B_{(k-1)m-1} + B_{m-1}\overline{\alpha}^{k-1}, \qquad (7)$$

which gives the generalization of (I.7.2).

Repeated application of (7) yields

$$B_{km-1} = B_{m-1}\overline{\alpha}^{k-1} + \alpha\big(B_{(k-1)m-1}\big)$$

$$= B_{m-1}\overline{\alpha}^{k-1} + B_{m-1}\alpha\overline{\alpha}^{k-2} + \alpha^2\big(B_{(k-2)m-1}\big)$$

$$= \cdots = B_{m-1}\big(\overline{\alpha}^{k-1} + \alpha\overline{\alpha}^{k-2} + \cdots + \alpha^{k-2}\overline{\alpha} + \alpha^{k-1}\big) = B_{m-1}\left(\frac{\alpha^k - \overline{\alpha}^k}{\alpha - \overline{\alpha}}\right)$$

as claimed.

For example, let $t = [\overline{1; 2, 3, 2, 1}]$, so $m = 5$, and we easily find the following table of values:

k	-1	0	1	2	3	4
a_k		1	2	3	2	1
A_k	1	1	3	10	23	33
B_k	0	1	2	7	16	23

Formula (I.2.5) becomes

$$t = \frac{33t + 23}{23t + 16} \quad \text{so} \quad 23t^2 - 17t - 23 = 0$$

and then

$$t = \frac{17 + \sqrt{2405}}{46},$$

$$\alpha = 23\left(\frac{17 + \sqrt{2405}}{46}\right) + 16 = \frac{49 + \sqrt{2405}}{2}$$

and

$$B_{5k-1} \;=\; \frac{23}{\sqrt{2405}}\left(\left(\frac{49+\sqrt{2405}}{2}\right)^{k}-\left(\frac{49-\sqrt{2405}}{2}\right)^{k}\right).$$

Thus $B_4 = 23$, $B_9 = 1127$, $B_{14} = 55246$, $B_{19} = 2708181$, ... Since $B_{(k+1)m-1} = A_{m-1}B_{km-1} + A_{m-2}B_{km-2}$, we also can find that

$$B_{13} \;=\; \frac{2708181 - 33\cdot 55246}{23} \;=\; 38481$$

and then the values of $B_{15}, \ldots , B_{18}$ can be recovered from B_{13}, B_{14} and the recurrence formulas.

NOTES

§1. See also Hardy and Wright [1971], LeVeque [1977], Niven and Zuckerman [1980], and Perron [1954].

§2. The material of this section also appears as an exercise in LeVeque [1977]. The method is essentially that of Hickerson [1973], in which the estimate $L(d) = \mathcal{O}(d^{\epsilon + 1/2})$ is obtained. It is clear from the proof that slightly more is true because the square factors of Q may be disregarded (see Podsypanin [1982]).

§3. The linear recurrence (1) is really a special case of the "second order linear homogeneous recurrence" given by $x_{k+1} = a_{k-1}x_k + b_{k-1}x_{k-1}$, which is best studied by way of the general continued fraction (see §1 of Perron [1954])

$$a_0 \;+\; \cfrac{b_1}{a_1 + \cfrac{b_2}{a_2 + \cdots}}$$

with convergents A_k/B_k obeying the recurrences $A_{k+1} = a_{k+1}A_k + b_{k+1}A_{k-1}$ and $B_{k+1} = a_{k+1}B_k + b_{k+1}B_{k-1}$ instead of (I.1.2). The particular case when the a_k's and b_k's are constants a and b is well known (see Niven and Zuckerman [1980]), and may be solved using the roots of the "characteristic equation" $x^2 = ax + b$. Higher order linear homogeneous recurrences with constant coefficients can also be solved using power series (see Hardy [1967]).

Chapter IV

APPLICATIONS

§1. Gear ratio problems.

The first treatment of continued fractions as best approximations was given by Christiaan Huygens as auxiliary material in his design of a gear driven model of the solar system. Although remembered as the Dutch physicist and astronomer who developed a wave theory of light, invented the first accurate pendulum clock and discovered the rings of Saturn, Huygens merits a footnote in number theory for his analysis of continued fractions and their application to approximation problems. His "automatic planetarium" was designed to show the relative motions of the six known planets around the sun (and, separately, the moon around the earth) and he took great pains to make it as accurate as possible. Each model planet was to be attached to a large circular gear which was then driven by a smaller gear mounted on a common drive shaft. Using the best data available in 1680 and approximating the year as $365\frac{35}{144}$ days, Huygens calculated the ratios of the periods of the planet's orbits to the earth's year and obtained the following values:

Mercury	Venus	Earth	Mars	Jupiter	Saturn
$\frac{25335}{105190}$	$\frac{64725}{105190}$	1	$\frac{197836}{105190}$	$\frac{1247057}{105190}$	$\frac{3095277}{105190}$

After expanding these ratios as continued fractions, he chose gear pairs corresponding to convergents and such that the component numbers would be practical to construct as gear teeth.

planet	Mercury	Venus	Mars	Jupiter	Saturn
convergent	$\frac{A_5}{B_5}=\frac{33}{137}$	$\frac{A_5}{B_5}=\frac{8}{13}$	$\frac{A_5}{B_5}=\frac{79}{42}$	$\frac{A_3}{B_3}=\frac{83}{7}$	$\frac{A_1}{B_1}=\frac{59}{2}$
gear teeth	33:137	32:52	158:84	166:14	118:4

59

Huygens gave a careful account of the errors resulting from his choices (he knew that the convergents approximate alternately from below and above) and his manuscripts from 1680-82 show continued recalculation to find better ratios. He restudied his data and approximations for Saturn several times, revised the ratio to $\frac{77708431}{2640858}$ and ultimately decided to use 206:7 (the third convergent) in spite of the difficulty in cutting a gear with more than two hundred teeth. He also made the happy observation that for Mercury he could factor the ninth convergent $\frac{204}{847}$ as $\frac{12 \cdot 17}{7 \cdot 121}$ and then realize this ratio as a gear train consisting of four gears with teeth 12:7,17:121 and the middle gears on a common shaft.

Closely related to the approximation problem represented by Huygens's model is the construction of a calendar that accurately determines the seasons (which depend on the revolution of the earth around the sun) by counting the days (which depend on the rotation of the earth about its axis). The Julian calendar used the approximation 1 year $\doteq$ $365\frac{1}{4}$ days and was implemented by extending every fourth 365 day "common" year by one extra day to form a "leap" year. After sixteen centuries of use, the disagreement between the Julian calendar and the actual motion of the earth around the sun amounted to ten days. In 1582, Pope Gregory XIII proclaimed a revised calendar based on the approximation 1 year $\doteq$ $365\frac{97}{400}$ days and implemented by omitting one leap year every century except every fourth century. Since this correction to the Julian calendar was chosen to occur in years divisible by 100, the Gregorian calendar is more accurate yet simple to use.

The "international standard system of units" defines the basic unit of time to be the ephemeris second, which is 1/31556925.9747 of the tropical year from 1900 January $0^d 12^h$ ephemeris time. Since one day is 84600 seconds by definition, the tropical year is

$$\frac{315569259747}{864000000} = [365; 4,7,1,3,5,6,1,1,3,1,7,7,1,1,1,1,2]$$

days long and the construction of a calendar reduces to selecting an appropriate approximation of the error

$$c = \frac{7750361}{32000000} = [0; 4,7,1,3,5,6,1,1,3,1,7,7,1,1,1,1,2]$$

between the tropical year and the common year. The first few convergents of c are

k	0	1	2	3	4	5
$\dfrac{A_k}{B_k}$	$\dfrac{0}{1}$	$\dfrac{1}{4}$	$\dfrac{7}{29}$	$\dfrac{8}{33}$	$\dfrac{31}{128}$	$\dfrac{163}{673}$

and so the Julian calendar is just a realization of the first convergent. If our notation system was based on powers of 2 instead of powers of 10, we could have from the fourth convergent a calendar in which leap years occur every fourth year with every thirty-second leap year omitted; the annual error would then be

$$c - \frac{31}{128} \doteq 0.00001128,$$

which amounts to the loss of one day every hundred-thousand years.

To construct a calendar based on a cycle b different from the denominator of a convergent of c, we must construct the number a of leap years to include during that cycle and we wish to have $|bc - a|$ as small as possible. We know from Chapter II that given a real number t and a positive integer b smaller than some convergent denominator B_{N+1} of t, there is a unique Ostrowski representation

$$b = \sum_{k=0}^{N} c_{k+1} B_k$$

and the number

$$A = A(b) = \sum_{k=0}^{N} c_{k+1} A_k$$

satisfies

$$|bt - A| < 1$$

by the proof of Theorem II.4.1. Thus the integer

$$a = a(b) = \begin{cases} A & \text{if } |bt - A| \leq 1/2 \\ A+1 & \text{if } bt - A > 1/2 \\ A-1 & \text{otherwise} \end{cases} \tag{1}$$

satisfies $|bt - a| \leq 1/2$ and will solve our construction problem. Notice that these calculations may be carried out using only integer arithmetic since, by Theorem

II.3.1, any convergent A_k/B_k with $k \geq N+1$ may be used in place of t without changing any of our inequalities. In the case of the Gregorian calendar, $b = 400$ becomes

$$b = 3 \cdot B_4 + 0 \cdot B_3 + 0 \cdot B_2 + 4 \cdot B_1 + 0 \cdot B_0$$

and then

$$A = 3 \cdot A_4 + 4 \cdot A_1 = 3 \cdot 31 + 4 \cdot 1 = 97$$

satisfies $|bc - A| < 1/2$, so $a = 97$ as expected. The resulting error is

$$c - \frac{97}{400} \doteq -0.00030122,$$

which is a gain of three days every ten-thousand years. Applying the same method to $b = 300$, we have

$$b = 2 \cdot B_4 + 1 \cdot B_3 + 0 \cdot B_2 + 2 \cdot B_1 + 3 \cdot B_0,$$
$$A = 2 \cdot 31 + 1 \cdot 8 + 2 \cdot 1 + 3 \cdot 0 = 72,$$
$$bc - A \doteq 0.6596 > 1/2 \quad \text{so} \quad a = A + 1 = 73$$

and an error of

$$c - \frac{73}{300} \doteq -0.00113455,$$

while $b = 500$ gives

$$b = 3 \cdot B_4 + 3 \cdot B_3 + 0 \cdot B_2 + 4 \cdot B_1 + 1 \cdot B_0,$$
$$A = 3 \cdot 31 + 3 \cdot 8 + 4 \cdot 1 + 1 \cdot 0 = 121,$$
$$|bc - A| < 1/2 \quad \text{and so} \quad a = 121.$$

It is interesting that these last calculations give a calendar with the extra leap year at the end of every fifth century instead of every fourth as in the Gregorian calendar and having a smaller error:

$$c - \frac{121}{500} \doteq 0.000198781,$$

which is a loss of two days every ten-thousand years.

While Huygens could have his gear ratios realized with custom made gears, the post-industrial revolution version of his problem more closely resembles the calendar problem: from a selection of standard gears, how can a particular real number be well approximated? For instance, given a supply of gears having from 20 to 120 teeth, can we redesign Huygens's planetarium with at least the same accuracy? If we restrict ourselves to gear trains with four gears of the type Huygens used for Mercury, for each of his five ratios r we must find four integers n_1, n_2, n_3 and n_4 such that

$$20 \leq n_1, n_2, n_3, n_4 \leq 120 \quad \text{and} \quad \left| r - \frac{n_1 n_3}{n_2 n_4} \right| \text{ is minimized.}$$

Since we know the value of each r to greater precision than we can achieve with the given gears, we may apply our method for the calendar problem. We begin by making an ordered list $\mathcal{L}$ of the possible values of $n \cdot m$, where $20 \leq n, m \leq 120$, so that $\mathcal{L}$ contains all possible choices for the numerators $n_1 \cdot n_3$ and the denominators $n_2 \cdot n_4$ (in this example, $\mathcal{L}$ will contain 3479 different numbers since some of them represent several possible pairs of gears). For each b in $\mathcal{L}$, we find the Ostrowski representation of b relative to r and construct $a = a(b)$ as in (1). If a is in $\mathcal{L}$, we set $a' = a$ while if a is not in $\mathcal{L}$, there are three possibilities: if $a > 120 \cdot 120$, we choose $a' = 120 \cdot 120$; if $a < 20 \cdot 20$, we choose $a' = 20 \cdot 20$; and if $20 \cdot 20 < a < 120 \cdot 120$, we can find the greatest lower bound a_{lower} in $\mathcal{L}$ of a and the least upper bound a_{upper} in $\mathcal{L}$ of a and then choose a' to be the nearest one to br. After finding a' for one b in $\mathcal{L}$, we then work our way through the rest of $\mathcal{L}$, revising our "best" pair of values for b and a' as we go along. After carrying out this program for Huygens's ratios, we obtain the following gear choices:

planet	Mercury	Venus	Mars	Jupiter	Saturn
ratio r	$\frac{5067}{21038}$	$\frac{12945}{21038}$	$\frac{98918}{52595}$	$\frac{1247057}{105190}$	$\frac{77708431}{2640858}$
gear train $(n_1 n_3 / n_2 n_4)$	$\frac{37 \cdot 53}{69 \cdot 118}$	$\frac{38 \cdot 107}{56 \cdot 118}$	$\frac{97 \cdot 113}{62 \cdot 94}$	$\frac{106 \cdot 119}{28 \cdot 38}$	$\frac{107 \cdot 110}{20 \cdot 20}$

We note that these gear trains all give closer approximations than those selected by

Huygens, even with his factored convergent for Mercury. Of course, some of our solutions are not unique; for instance, we also could use $(106 \cdot 85)/(20 \cdot 38)$ for Jupiter.

It is worth remarking that the above method is completely general and can be carried out using only integer multiplications and comparisons. We have thus given an integer algorithm that solves the "general gear ratio problem"

$$\min_{a \in N, b \in \mathfrak{D}} |t - a/b|$$

where t is a positive real number and N and $\mathfrak{D}$ are given finite sets of positive integers from which the numerators a and denominators b are selected.

§2. Pell's equation.

The equation $x^2 - dy^2 = 1$ was connected with the name of John Pell by Euler in a letter (10 August 1730) to Goldbach because Euler believed Pell responsible for a solution technique (actually due to Brouncker) that appeared in a book by Wallis. However, priority seems to fall to Fermat as he discussed the equation in 1657 and asked for a general method for finding solutions. We will study the integer solutions of the more general equation

$$x^2 - dy^2 = N, \tag{1}$$

where N is a given integer and the integer $d > 1$ is not a perfect square. It is clear that we may take x and y to be positive integers.

If $N > 0$, we may factor (1) as a product of positive factors

$$(x - y\sqrt{d})(x + y\sqrt{d}) = N$$

to obtain

$$(x - y\sqrt{d})^2 + (x - y\sqrt{d})(2y\sqrt{d}) = N$$

and hence

$$0 < y(x - y\sqrt{d}) < \frac{N}{2\sqrt{d}}.$$

If $N < 0$, we can rewrite (1) as $y^2 - (1/d)x^2 = -N/d$ to reverse the roles of x and y and to obtain an equation with a positive quantity on the right side. Proceeding as before, we find that

$$\left(y - x\frac{1}{\sqrt{d}}\right)^2 + \left(y - x\frac{1}{\sqrt{d}}\right)\left(2x\frac{1}{\sqrt{d}}\right) = -N/d$$

and thus

$$0 < x\left(y - x\frac{1}{\sqrt{d}}\right) < \frac{-N}{2\sqrt{d}}.$$

When $0 < N < \sqrt{d}$, we see that $0 < y(x - y\sqrt{d}) < 1/2$ and thus $x - y\sqrt{d}$ is the same as $\| y\sqrt{d} \|$. Similarly, if $-\sqrt{d} < N < 0$, we have that $y - x/\sqrt{d}$ is the same as $\| x/\sqrt{d} \|$. Applying our results from §5 of Chapter II on the approximation $b \| bt \|$ $< 1/2$ together with the relationship from §5 of Chapter I between the convergents of $\sqrt{d}$ and $1/\sqrt{d}$, we immediately obtain:

Theorem 1. *If* $|N| < \sqrt{d}$, *then the relatively prime solutions of* (1) *are* $x = A_k$ *and* $y = B_k$, *where* A_k/B_k *is a convergent of* $\sqrt{d}$.

But which values of N can be so represented and will all such choices of x and y result in an N with $|N| < \sqrt{d}$? Let us write $\zeta_0 = \sqrt{d}$ and $\zeta_k = (P_k + \sqrt{d})/Q_k$ as in the second proof of Theorem III.1.2. By (I.2.5) we have

$$\sqrt{d} = \frac{A_k(P_{k+1} + \sqrt{d}) + A_{k-1}Q_{k+1}}{B_k(P_{k+1} + \sqrt{d}) + B_{k-1}Q_{k+1}} \tag{2}$$

and so

$$A_k = B_k P_{k+1} + B_{k-1}Q_{k+1} \quad \text{and} \quad dB_k = A_k P_{k+1} + A_{k-1}Q_{k+1},$$

by comparing the integer and the irrational parts of (2). Thus

$$A_k^2 - dB_k^2 = A_k(B_k P_{k+1} + B_{k-1}Q_{k+1}) - B_k(A_k P_{k+1} + A_{k-1}Q_{k+1})$$

$$= (A_k B_{k-1} - A_{k-1}B_k)Q_{k+1} = (-1)^{k+1}Q_{k+1},$$

and we find (as was known to Euler) that the value of $x^2 - dy^2$ for $x = A_k$ and $y = B_k$ depends only on the parity of k and the value of Q_{k+1}. Since the integers Q_k satisfy $\sqrt{d} - P_k < Q_k < \sqrt{d} + P_k$, the convergents may give values of $x^2 - dy^2$ larger than $\sqrt{d}$, but they can not give values larger than $2\sqrt{d}$. From Theorem III.1.6, $Q_k = 1$ only at the end of each period in the continued fraction of $\sqrt{d}$ and we have established:

Theorem 2. *If $N = 1$, then (1) is always solvable. If $N = -1$, then (1) is solvable only if the length of the period of $\sqrt{d}$ is odd.*

Returning to the examples of $\sqrt{d}$ given in §1 of Chapter III, we now see that both $x^2 - 29y^2 = 1$ and $x^2 - 29y^2 = -1$ are solvable, since the periodic block of the continued fraction for $\sqrt{29}$ has odd length, while $x^2 - 34y^2 = 1$ but not $x^2 - 34y^2 = -1$ is solvable, since $\sqrt{34}$ has an even length periodic block. Further, from the values of the Q_k's, we see that the equations $x^2 - 29y^2 = \pm 4, \pm 5$ are solvable, but that $x^2 - 29y^2 = \pm 2, \pm 3$ are not.

We now consider $|N| > \sqrt{d}$. If (1) has a solution (x_0, y_0), then it will have infinitely many solutions: given a solution (X, Y) of $x^2 - dy^2 = 1$, the numbers

$$x_1 = x_0 X + y_0 Y \quad \text{and} \quad y_1 = x_0 Y + y_0 X$$

are also solutions of (1) since

$$x_1^2 - dy_1^2 = \left(x_0 X + y_0 Y\right)^2 - d\left(x_0 Y + y_0 X\right)^2 = x_0^2\left(X^2 - dY^2\right) - dy_0^2\left(X^2 - dY^2\right) = N.$$

Further, if (1) has a solution, then for $N > 0$ the quantity $x - y\sqrt{d}$ satisfies

$$0 < x - y\sqrt{d} = \frac{N}{x + y\sqrt{d}}$$

and can be made arbitrarily small; similarly, for $N < 0$ the quantity $y - x/\sqrt{d}$ satisfies

$$0 < y - x/\sqrt{d} = \frac{-N/d}{y + x/\sqrt{d}}$$

and can be arbitrarily small. Thus we again have reduced the study of (1) to the approximation problems $y \| y\sqrt{d} \| < N/2\sqrt{d}$ (when $N > 0$) and $x \| x/\sqrt{d} \| < -N/2\sqrt{d}$ (when $N < 0$).

If we require that $|N| < 2\sqrt{d}$ then, as for Theorem 1, we may apply our results from §5 of Chapter II on the approximation $b \| bt \| < 1$ to find that the relatively prime solutions of (1) now must be of the form $x = A_k$ and $y = B_k$ or $x = A_k + c_{k+2}A_{k+1}$ and $y = B_k + c_{k+2}B_{k+1}$. Since

$$x^2 - dy^2 = (y\sqrt{d} - x)(\overline{y\sqrt{d} - x})$$

is $D_k\overline{D}_k$ for $x = A_k$ and $y = B_k$, we have that

$$D_k\overline{D}_k = (-1)^{k+1}Q_{k+1}$$

(which also may be shown directly by induction). Thus the "intermediate convergent" $x = A_k + c_{k+2}A_{k+1}$ and $y = B_k + c_{k+2}B_{k+1}$ (where $1 \le c_{k+2} \le a_{k+2} - 1$) gives the value

$$(D_k + c_{k+2}D_{k+1})(\overline{D_k + c_{k+2}D_{k+1}}) = D_k\overline{D}_k\left(1 - \frac{c_{k+2}}{\zeta_{k+2}}\right)\left(1 - \frac{c_{k+2}}{\overline{\zeta}_{k+2}}\right)$$

$$= (-1)^{k+1}Q_{k+1}\left(1 - \frac{c_{k+2}}{\frac{P_{k+2} + \sqrt{d}}{Q_{k+2}}}\right)\left(1 - \frac{c_{k+2}}{\frac{P_{k+2} - \sqrt{d}}{Q_{k+2}}}\right)$$

$$= (-1)^{k+1}(Q_{k+1} + 2c_{k+2}P_{k+2} - c_{k+2}^2 Q_{k+2}).$$

Since the largest value of $Q_{k+1} + 2c_{k+2}P_{k+2} - c_{k+2}^2 Q_{k+2}$ occurs when $c_{k+2} = \| P_{k+2}/Q_{k+2} \|$ and is approximately $Q_{k+1} + P_{k+2}^2/Q_{k+2} = d/Q_{k+2}$, the intermediate convergents can give values as large as d (because $Q_{k+2} = 1$ at the end of the periodic block). From the continued fraction for $\sqrt{77}$ given in §1 of Chapter III, we see that $x^2 - 77y^2$ takes the four values 1, 4, -7, -13 for the convergents $x = A_k$ and $y = B_k$ (notice that $13 > \sqrt{77}$) and the eleven values 11, -17, -19, -28, -41, -52, -61, -68, -73, -76 and -77 for the intermediate convergents. As would be expected, -77 is attained only for $x = A_{4+6k} + 8A_{5+6k}$ and $y = B_{4+6k} + 8B_{5+6k}$, where $k \ge 0$.

It should be clear from Theorem II.5.3 that as the bound on $|N|$ increases, longer and longer combinations of the A_k's and B_k's will be required; but it also must be that expressions of a certain length can not give values below a corresponding size. More precisely, we now show:

Theorem 3. *The positive integer solutions of* $x^2 - dy^2 = N$, *where* $2K_1\sqrt{d} < |N|$ $< 2K_2\sqrt{d}$ *and* $K_1 < K_2$ *are positive constants, have the form*

$$
\begin{aligned}
x &= c_{n+1}A_n + c_{n+2}A_{n+1} + \cdots + c_{n+m}A_{n+m-1} \\
y &= c_{n+1}B_n + c_{n+2}B_{n+1} + \cdots + c_{n+m}B_{n+m-1}
\end{aligned}
\tag{3}
$$

(where, as usual, the coefficients c_{k+1} *satisfy* $0 \le c_1 < a_1$, $0 \le c_{k+1} \le a_{k+1}$ *for* $k \ge 0$, *and* $c_k = 0$ *if* $c_{k+1} = a_{k+1}$*) with length* m *such that*

$$
\tfrac{1}{2}\left(\log_{(1+2\sqrt{d})}(K_1) - 1 \right) < m < \tfrac{1}{2}\left(\log_g(2K_2 + 1) + 3 \right).
$$

Proof. The upper estimate for m follows from Theorem II.5.3, since $y \, \| \, y\sqrt{d} \, \|$ $< N/(2\sqrt{d}) < K_2$. From the upper estimate of Lemma II.4.1, $\| \, y\sqrt{d} \, \|$ is largest for integers of the form

$$
y = a_{n+1}B_n + a_{n+3}B_{n+2} + \cdots + a_{n+2m+1}B_{n+2m} = B_{n+2m+1} - B_{n-1},
$$

and for such numbers we have

$$
y \, \| \, y\sqrt{d} \, \| \; < \; y \, | \, a_{n+1}D_n - D_{n+1} | \; = \; y \, | -D_{n-1} | \; < \; \frac{B_{n+2m+1}}{B_n}.
$$

Since the partial quotients of $\sqrt{d}$ are bounded above by $2\sqrt{d}$, we have the estimate $B_{k+1}/B_k < 1 + 2\sqrt{d}$ and thus $y \, \| \, y\sqrt{d} \, \| \; < (1 + 2\sqrt{d})^{2m+1}$.

Of course, the estimate $B_{k+1}/B_k < 1 + 2\sqrt{d}$ is much too large for most of the partial quotients. For $d = 77$, there are fifty-five values of $|x^2 - 77y^2|$, ranging from 23 to 1156, for x and y as in (3) with length $m = 3$, and some of these values duplicate those from shorter expressions. For example, $x = 2A_1 + A_2$ $= 2 \cdot 9 + 35 = 53$ and $y = 2B_1 + B_2 = 2 \cdot 1 + 4 = 6$ gives

$$(2A_1 + A_2)^2 - 77(2B_1 + B_2)^2 \;=\; 53^2 - 77 \cdot 6^2 \;=\; 37$$

as does $x = A_1 + A_2 + A_3 = 9 + 35 + 79 = 123$ and $y = B_1 + B_2 + B_3 = 1 + 4 + 9 = 14$, since

$$(A_1 + A_2 + A_3)^2 - 77(B_1 + B_2 + B_3)^2 \;=\; 123^2 - 77 \cdot 14^2 \;=\; 37.$$

Since the period length of $\sqrt{77}$ is 6, we also have

$$(2A_{1+6k} + A_{2+6k})^2 - 77(2B_{1+6k} + B_{2+6k})^2 \;=\; 37$$

and

$$(A_{1+6k} + A_{2+6k} + A_{3+6k})^2 - 77(B_{1+6k} + B_{2+6k} + B_{3+6k})^2 \;=\; 37$$

for $k \geq 0$.

§3. Fermat's theorem on the sum of two squares.

In 1625, Albert Girard found that every number expressible as the sum of two integral squares was either a square, a prime p of the form $p = 4k + 1$, a product formed from such numbers, or twice one of the previous forms. Fermat considered the same problem and it is probable that he could prove his results, although the first proofs on record are due to Euler. Fermat regarded the fact that every prime of the form $4k + 1$ is a sum of two integral squares as the "fundamental theorem of right triangles" and it is on this result that we concentrate. While proofs of this theorem (using the language of quadratic residues) may be found in many number theory textbooks, the actual construction of integers x and y with $x^2 + y^2 = p$ is another matter. Essentially four different constructions are known and we shall present the earliest, due to Legendre in 1808, and a later construction, by Charles Hermite in 1848.

Let p be any odd prime and consider Pell's equation

$$x^2 - py^2 \;=\; 1. \tag{1}$$

As we saw in §2, such an equation always has a solution, so let X and Y be the smallest positive solutions. We write $Y = abc$, where a, b, $c > 0$ are integers, and consider $(X+1)(X-1) = pY^2$. Either

$$\begin{cases} X+1 = ab^2p \\ X-1 = ac^2 \end{cases} \text{ so that } \frac{-2}{a} = c^2 - b^2p$$

or

$$\begin{cases} X+1 = ab^2 \\ X-1 = ac^2p \end{cases} \text{ so that } \frac{2}{a} = b^2 - c^2p. \tag{2}$$

Thus a can be only 1 or 2. We can exclude the possibility that $a = 2$ in (2), since we supposed that X and Y were the smallest positive solutions of (1). There are three equations to consider further:

$$\text{(a) } c^2 - pb^2 = -1, \text{ (b) } c^2 - pb^2 = -2 \text{ and (c) } b^2 - pc^2 = 2. \tag{3}$$

Since $p > 2$ can be only of the form $4k+1$ or $4k+3$, we may try various combinations of odd and even choices of b and c in the equations (3). For example, if $p = 4k+1$ and $b = 2M$ and $c = 2N$ are even, equation (3b) would become

$$4N^2 - (4k+1)(4M^2) = -2,$$

which is impossible since -2 is not an integer multiple of 4. In this way we can see that $x^2 - py^2$ is never -1 if $p = 4k+3$ and that $x^2 - py^2$ is never 2 or -2 if $p = 4k+1$. Thus $p = 4k+1$ if and only if $x^2 - py^2 = -1$ has a solution. But then from our study of Pell's equation, $p = 4k+1$ if and only if the continued fraction for $\sqrt{p}$ has a odd period length; thus, by (III.1.7), we have that

$$\sqrt{p} = [a_0; \overline{a_1, \ldots, a_n, a_n, \ldots, a_1, 2a_0}],$$

where $m = 2n+1$. Moreover, ζ_{n+1} is purely periodic and by Theorem III.1.4 its conjugate η_{n+1} has $-1/\eta_{n+1} = \zeta_{n+1}$. So $\zeta_{n+1} \cdot \eta_{n+1} = -1$. That is,

$$\frac{P_{n+1} + \sqrt{p}}{Q_{n+1}} \cdot \frac{P_{n+1} - \sqrt{p}}{Q_{n+1}} = -1,$$

which means

$$p = P_{n+1}^2 + Q_{n+1}^2,$$

and the construction is complete.

Using the examples of §1 in Chapter III, we have that $29 = P_3^2 + Q_3^2 = 2^2 + 5^2$. Although we have proven the construction only for primes of the form $4k + 1$, we remark that any integer $D > 0$ such that $\sqrt{D}$ has an odd periodic block length $m = 2n + 1$ can be represented as $D = P_{n+1}^2 + Q_{n+1}^2$; for example, $493 = P_5^2 + Q_5^2 = 18^2 + 13^2$. However, it is not the case that if D can be written as the sum of two squares that then $\sqrt{D}$ has an odd length periodic block; for instance, $34 = 3^2 + 5^2$ yet $\sqrt{34}$ has an even length period.

In the language of quadratic residues, Legendre's construction not only shows that -1 is a residue for primes of the form $4k + 1$ and is a non-residue for primes of the form $4k + 3$, but also solves the congruence $x^2 \equiv -1 \pmod{p}$ with a square multiple of p. Our second construction requires only a solution of the congruence.

Let $x^2 \equiv -1 \pmod{p}$ and consider the continued fraction of x/p. Then there is a unique integer n with $B_n < \sqrt{p} < B_{n+1}$. Since x/p is between A_n/B_n and A_{n+1}/B_{n+1},

$$\frac{x}{p} = \frac{A_n}{B_n} + \frac{\epsilon}{B_n B_{n+1}},$$

where $|\epsilon| < 1$. Then $x B_n - p A_n = \epsilon p / B_{n+1}$ and, since $B_{n+1} > \sqrt{p}$, we see that $(x B_n - p A_n)^2 < p$. Thus

$$0 < (x B_n - p A_n)^2 + B_n^2 < 2p.$$

But $(x B_n - p A_n)^2 + B_n^2 = B_n^2(x^2 + 1) + p \cdot (p A_n^2 - 2x A_n B_n)$ is divisible by p, since $x^2 + 1 \equiv 0 \pmod{p}$, and consequently must be the desired representation of p.

To apply Hermite's construction in the case $p = 29$, we must first solve $x^2 \equiv -1 \bmod 29$. From our discussion of Pell's equation in §2, we know that $A_k^2 - 29 B_k^2 = (-1)^{k+1} Q_{k+1}$. Thus $A_4^2 - 29 B_4^2 = (-1)^5 \cdot 1$ and so $x = A_4 \equiv 12 \bmod 29$ will do. Alternatively, given any non-residue y of p, we may take $x \equiv y^{(p-1)/4}$

mod p. Since 2 is a non-residue of primes of the form $8k + 5$, this becomes $x \equiv 2^7$ mod 29, which also gives $x = 12$. Expanding 12/29 as a continued fraction, we find $12/29 = [0;\ 2,\ 2,\ 2,\ 2]$ with convergents 0, 1/2, 2/5, 5/12, 12/29. So $B_2 < \sqrt{29} < B_3$, $n = 2$ and we have $29 = (12B_2 - 29A_2)^2 + B_2^2 = (12 \cdot 5 - 29 \cdot 2)^2 + (5)^2 = 2^2 + 5^2$. Since also $17^2 \equiv -1$ mod 29, we may repeat the construction using $17/29 = [0;\ 1,\ 1,\ 2,\ 2,\ 2]$.

§4. Hall's Theorem.

Numbers represented by continued fractions with bounded digits have many similarities to those expressible with restricted digits in a radix representation system. In this section, we will describe some properties of the Cantor set obtained from the unit interval by removal of "middle thirds" and obtain similar results for sets of continued fractions with partial quotients no greater than a given positive integer.

The *Cantor set* $\mathcal{C}$ is a subset of the closed interval $[0, 1]$ constructed as follows. Let $\mathcal{C}_0 = [0, 1]$ and remove the open interval $(1/3, 2/3)$

to obtain $\mathcal{C}_1 = [0, 1/3] \cup [2/3, 1]$. Similarly, $\mathcal{C}_{k+1}$ is obtained from $\mathcal{C}_k$ by removing the open middle third from each of the closed intervals comprising $\mathcal{C}_k$. Since only open intervals are removed, each $\mathcal{C}_k$ is closed and the limit set $\mathcal{C}$ is also closed. Further, each point of $\mathcal{C}$ clearly is a limit point of $\mathcal{C}$ and so $\mathcal{C}$ is a "perfect" set. Since the length of the intervals forming $\mathcal{C}_{k+1}$ is 2/3's of the length of those forming $\mathcal{C}_k$, $\mathcal{C}$ contains no interval and is a set of "measure zero."

Writing the number $t \in \mathcal{C}_0$ in "ternary" notation,

$$t = 0 + \frac{d_1}{3} + \frac{d_2}{9} + \frac{d_3}{27} + \cdots = 0.d_1 d_2 d_3 \ldots,$$

where each of the digits d_k is $0 \leq d_k \leq 2$, we see that $\mathcal{C}_1$ consists of those numbers having $d_1 \neq 1$ (although $1/3 = 0.100 \ldots$, it is also $0.0222 \ldots$), $\mathcal{C}_2$ has d_1 and $d_2 \neq 1$

and so on. Thus C may also be regarded as those numbers in C_0 whose ternary expansions consist only of the digits 0 and 2.

In spite of the fact that C contains no interval, we can show that $C + C = C_0 + C_0$; that is, every $t \in [0, 2]$ can be expressed as a sum $t_1 + t_2$ with t_1, $t_2 \in C$. To see that $C_1 + C_1 = C_0 + C_0$, we remove one middle third at a time to first obtain $C_1 + C_0$:

$$C_1 \qquad + \qquad C_0$$

$$\begin{array}{ccccc} | & | & | & | \\ 0 & \frac{1}{3} & \frac{2}{3} & 1 \end{array} \qquad \begin{array}{cc} | & | \\ 0 & 1 \end{array}$$

$$= \qquad [0, \tfrac{4}{3}] \qquad \bigcup \qquad [\tfrac{2}{3}, 2] \qquad = C_0 + C_0$$

$$\begin{array}{cccc} | & | & | & | \\ 0 & \frac{4}{3} & \frac{2}{3} & 2 \end{array}$$

because $[0, 1/3] + C_0$ overlaps $[2/3, 1] + C_0$. In general, given two closed intervals $[a, b]$ and $[c, d]$ of lengths $|[a, b]| = b - a$ and $|[c, d]| = d - c$, respectively, we can remove (a', b') from $[a, b]$, where $a < a' < b' < b$, and consider the sums:

$$[a, b] \qquad + \qquad [c, d] \qquad = \qquad [a + c, b + d]$$

$$\begin{array}{cc} | & | \\ a & b \end{array} \qquad \begin{array}{cc} | & | \\ c & d \end{array} \qquad \begin{array}{cc} | & | \\ a+c & b+d \end{array}$$

and

$$\big([a, a'] \cup [b', b]\big) \qquad + \qquad [c, d] \qquad = \qquad [a + c, a' + d] \qquad \bigcup \qquad [b' + c, b + d]$$

$$\begin{array}{cccc} | & | & | & | \\ a & a' & b' & b \end{array} \qquad \begin{array}{cc} | & | \\ c & d \end{array} \qquad \begin{array}{cc} | & | \\ a+c & a'+d \end{array} \qquad \begin{array}{cc} | & | \\ b'+c & b+d \end{array}$$

These last two intervals overlap to form $[a + c, b + d]$ only if $b' + c \leq a' + d$; that is, only if $|[a', b']| \leq |[c, d]|$. Thus, if the gap removed from one interval is no larger than the other interval in the sum, the sum will remain the same. Returning to the Cantor set, to pass from $C_1 + C_0 = C_0 + C_0$ to $C_1 + C_1 = C_0 + C_0$, we note that the gap introduced in the second C_0 is no larger than either of the intervals making up C_1. Thus $[0, 1/3] + C_1$ is the same as $[0, 1/3] + C_0$ and similarly for $[2/3, 1] + C_1$ and

$[2/3, 1] + \mathcal{C}_0$, and this establishes that $\mathcal{C}_1 + \mathcal{C}_1 = \mathcal{C}_0 + \mathcal{C}_0$. This type of argument easily extends to yield the following general:

Lemma 1. *Let* I_0, I_1, $\ldots$, $I_n \subset I$ *be closed sub-intervals of a closed interval* I *such that*

$$\left(I_0 \cup \bigcup_{k=1}^{n} I_k\right) + \left(I_0 \cup \bigcup_{k=1}^{n} I_k\right) = I + I$$

and suppose that an open interval $G \subset I_0$ *is removed from* I_0 *to leave two closed intervals* L, $R \subset I_0$. *If* $|G| \leq |I_k|$ *for* $k = 1, 2, \ldots, n$ *then*

$$\left((L \cup R) \cup \bigcup_{k=1}^{n} I_k\right) + \left((L \cup R) \cup \bigcup_{k=1}^{n} I_k\right) = I + I.$$

This observation allows us to conclude that $\mathcal{C}_{k+1} + \mathcal{C}_{k+1} = \mathcal{C}_0 + \mathcal{C}_0$ given that $\mathcal{C}_k + \mathcal{C}_k = \mathcal{C}_0 + \mathcal{C}_0$, since each $\mathcal{C}_{k+1}$ may be obtained in turn by removing the 2^k gaps of length $1/3^{k+1}$ from $\mathcal{C}_k$ one at a time, and thus $\mathcal{C} + \mathcal{C} = \mathcal{C}_0 + \mathcal{C}_0$. It is important to note that this proof does not provide the decomposition of $t \in [0, 2]$ into $t_1 + t_2$ with $t_1, t_2 \in \mathcal{C}$ but only asserts that such a representation exists. Similar results for decimal representations with excluded digits are easy to obtain.

We now turn to continued fractions with restricted partial quotients. Let N be a given positive integer and let $\mathrm{CF}(N)$ denote the set of all continued fractions with $a_0 = 0$ and $1 \leq a_k \leq N$ for $k \geq 1$; that is,

$$\mathrm{CF}(N) = \left\{[0; a_1, a_2, \ldots] : 1 \leq a_k \leq N \text{ for } 1 \leq k\right\}.$$

Of course, if $N = 1$ then $\mathrm{CF}(N)$ consists of the single quadratic surd $(-1 + \sqrt{5})/2$, so let us now suppose that $N \geq 2$. If we also let

$$\mathrm{CF}(N)^*_{i+1} = \left\{[0; a_1, a_2, \ldots] : 1 \leq a_k \leq N \text{ for } 1 \leq k \leq i\right\}$$

for $i \geq 0$, then

$$\mathrm{CF}(N) = \bigcap_{i=0}^{\infty} \mathrm{CF}(N)^*_{i+1}$$

and each $\mathrm{CF}(N)^*_{i+1}$ is a finite union of closed intervals:

$$\text{CF}(N)^*_{i+1} = \bigcup_{\substack{\frac{A_i}{B_i} = [0; a_1, \ldots, a_i] \\ 1 \le a_k \le N}} \left\{ \frac{A_i \zeta_{i+1} + A_{i-1}}{B_i \zeta_{i+1} + B_{i-1}} : 1 \le \zeta_{i+1} \le N+1 \right\},$$

since $\zeta_{i+1} = N + 1$ corresponds to the continued fraction $[0; a_1, \ldots, a_i, N, 1]$. By (I.2.6),

$$\frac{A_i \zeta_{i+1} + A_{i-1}}{B_i \zeta_{i+1} + B_{i-1}} = \frac{A_i}{B_i} + \frac{(-1)^i}{B_i^2(\zeta_{i+1} + \xi_i)}$$

and so the continued fractions with $1 \le \zeta_{i+1} \le N + 1$ fill out an interval of length

$$\frac{1}{B_i^2} \left(\frac{1}{1 + \xi_i} - \frac{1}{N + 1 + \xi_i} \right)$$

within an interval of length

$$\frac{1}{B_i^2} \left(\frac{1}{1 + \xi_i} \right).$$

Thus the construction of $\text{CF}(N)^*_{i+1}$ from $\text{CF}(N)^*_i$ by the removal of open intervals reduces the length of $\text{CF}(N)^*_i$ by a factor of

$$1 - \frac{1 + \xi_i}{N + 1 + \xi_i} < 1 - \frac{1}{N+1}$$

(since $0 < \xi_i < 1$) and so $\text{CF}(N)$ is a set of measure zero.

In order to find an analogous result to our observation that the sum of the Cantor set with itself forms an interval, we shall use a slightly different construction for $\text{CF}(N)$. Since $\text{CF}(N)$ clearly consists of irrational numbers only and the rational endpoints $1/(N+1)$ and 1 of $\text{CF}(N)^*_1$ are lost when passing to $\text{CF}(N)^*_2$ (and similarly when passing from $\text{CF}(N)^*_i$ to $\text{CF}(N)^*_{i+1}$), we may instead study the sets

$$\text{CF}(N)_{i+1} = \left\{ [0; a_1, a_2, \ldots, a_i, \zeta_{i+1}] : \begin{array}{c} 1 \le a_k \le N \text{ for } 1 \le k \le i \\ \text{and} \\ [1; \overline{N, 1}] \le \zeta_{i+1} \le [N; \overline{1, N}] \end{array} \right\},$$

since we still have that

$$\mathrm{CF}(N) \;=\; \bigcap_{i=0}^{\infty} \mathrm{CF}(N)_{i+1}.$$

Now $\mathrm{CF}(N)_{i+2}$ is formed from $\mathrm{CF}(N)_{i+1}$ by deleting open intervals from the interiors of the closed intervals comprising $\mathrm{CF}(N)_{i+1}$ and the resulting sets are composed of closed intervals with no loss of isolated rational points. Thus

$$\mathrm{CF}(N)_1 \;=\; \left\{ \frac{A_0\zeta_1 + A_{-1}}{B_0\zeta_1 + B_{-1}} : a_0 = 0 \text{ and } [1; \overline{N, 1}] \le \zeta_1 \le [N; \overline{1, N}] \right\}$$

is a closed interval with endpoints $[0; \overline{N, 1}]$ and $[0; \overline{1, N}]$. $\mathrm{CF}(N)_2$ is obtained from $\mathrm{CF}(N)_1$ by deleting the $N-1$ open intervals corresponding to

$$[k; \overline{1, N}] \;<\; \zeta_1 \;<\; [k+1; \overline{N, 1}]$$

for $k = 1, 2, \ldots, N-1$ and so

$$\mathrm{CF}(N)_2 \;=\; \bigcup_{\substack{\frac{A_1}{B_1} = [0;\, a_1] \\ 1 \le a_1 \le N}} \left\{ \frac{A_1\zeta_2 + A_0}{B_1\zeta_2 + B_0} : [1; \overline{N, 1}] \le \zeta_2 \le [N; \overline{1, N}] \right\}$$

and, more generally, we set

$$\mathrm{CF}(N)_{i+1} \;=\; \bigcup_{\substack{\frac{A_i}{B_i} = [0;\, a_1, \ldots, a_i] \\ 1 \le a_1, \ldots, a_i \le N}} \left\{ \frac{A_i\zeta_{i+1} + A_{i-1}}{B_i\zeta_{i+1} + B_{i-1}} : [1; \overline{N, 1}] \le \zeta_{i+1} \le [N; \overline{1, N}] \right\}.$$

If we are to apply Lemma 1 to the sequence of closed sub-intervals created in passing from $\mathrm{CF}(N)_{i+1}$ to $\mathrm{CF}(N)_{i+2}$, we must examine the lengths of the gaps so generated. For any particular choice of

$$\frac{A_i}{B_i} \;=\; [0; a_1, \ldots, a_i],$$

where $1 \le a_1, \ldots, a_i \le N$, the remaining closed intervals on either side of a gap need to be at least as large as the remaining intervals when all the gaps have been removed. Thus we must check that both

$$\frac{1}{B_i^2}\left(\frac{1}{[k;\,\overline{1,\,N}]+\xi_i}-\frac{1}{[k+1;\,\overline{N,\,1}]+\xi_i}\right) \leq \frac{1}{B_i^2}\left(\frac{1}{[k;\,\overline{N,\,1}]+\xi_i}-\frac{1}{[k;\,\overline{1,\,N}]+\xi_i}\right)$$

and

$$\frac{1}{B_i^2}\left(\frac{1}{[k;\,\overline{1,\,N}]+\xi_i}-\frac{1}{[k+1;\,\overline{N,\,1}]+\xi_i}\right)$$

$$\leq \frac{1}{B_i^2}\left(\frac{1}{[k+1;\,\overline{N,\,1}]+\xi_i}-\frac{1}{[k+1;\,\overline{1,\,N}]+\xi_i}\right)$$

hold for $k = 1, 2, \ldots, N-1$.

Writing $\alpha = [0;\,\overline{N,\,1}] = (-N+\sqrt{N^2+4N})/2N$, so that $[0;\,\overline{1,\,N}] = N\alpha$, and $X = k+\alpha+\xi_i$, we need to show that

$$\frac{1}{X+(N-1)\alpha}-\frac{1}{X+1} \leq \frac{1}{X}-\frac{1}{X+(N-1)\alpha} \tag{1}$$

and

$$\frac{1}{X+(N-1)\alpha}-\frac{1}{X+1} \leq \frac{1}{X+1}-\frac{1}{X+1+(N-1)\alpha}. \tag{2}$$

But (1) is equivalent to

$$2X^2+2X \leq 2X^2+(1+2(N-1)\alpha)X+(N-1)\alpha$$

and this holds for all the k's since $2(N-1)\alpha > 1$. Writing $\beta = (N-1)\alpha$, (2) becomes

$$\frac{1}{X+\beta}+\frac{1}{X+\beta+1} \leq \frac{2}{X+1},$$

which holds provided

$$\frac{1-2\beta^2}{2\beta-1} \leq X.$$

Since $X > 1$ can be very close to 1, let us look at the inequality with X replaced by 1; that is, $0 < \beta^2+\beta-1$ and so we need $\beta > 1/g$, where $g = (1+\sqrt{5})/2$ as in §7 of Chapter I. Using the relation $g^2 = 1+g$, we solve

$$\frac{N-1}{2N}(-N+\sqrt{N^2+4N}) > \frac{1}{g}$$

to find

$$N > \frac{(3+g)+\sqrt{3(2+g)}}{2} \doteq 3.956$$

and thus the integer N must be $N \geq 4$. We are thus led to:

Theorem 1. $CF(N) + CF(N) = [(-N + \sqrt{N^2 + 4N})/N, \ -N + \sqrt{N^2 + 4N}]$ *if* $N \geq 4$.

Proof. Although we have shown that the gaps removed from each closed interval making up $CF(N)_{i+1}$ are shorter than the two remaining closed intervals to the left and right of each gap, it is clear that some of the gaps to be removed from one part of $CF(N)_{i+1}$ may be longer than the smallest interval contained in another part of $CF(N)_{i+1}$. In order to apply Lemma 1, let us list the gaps G_k to be removed from $CF(N)_1$ to obtain $CF(N)$ in order of decreasing length and let them be removed from $CF(N)_1$ in this order. That is, we consider the sequence of sets of closed intervals $S_0 = CF(N)_1$, $S_1 = S_0 - G_1$, ... , $S_{k+1} = S_k - G_{k+1}$, ... Clearly, $\lim S_k = CF(N)$ and if we can show that Lemma 1 applies to S_k, we shall have $S_{k+1} + S_{k+1} = S_0 + S_0$ at each step and the proof will be finished.

Let G_k be removed from the closed interval I contained in S_k and let L and R be the remaining parts of I to the left and right of G_k. If R equals or contains an interval of the form discussed in (1) or (2) then $|R| \geq |G_k|$. If not, then R abuts (on the side opposite from G_k) an interval, say G_{k*}, that has already been removed from an interval equal to or containing an interval of the form discussed in (1) and (2). So then $|G_{k*}| \leq |R|$ and, since G_{k*} was removed before G_k, we must have $|G_k| \leq |G_{k*}|$ so that $|G_k| \leq |R|$. Similarly for L and the proof is complete.

The case $N = 4$ of this theorem was proven by Hall [1947] and he used it to deduce the following result, which will be of great importance in our later discussion of the Markov spectrum.

Corollary ("Hall's Theorem"). *Any real number t can be expressed as*

$$t = n + [0; a_1, a_2, \dots] + [0; a_1^*, a_2^*, \dots],$$

where n is an integer and the partial quotients satisfy $1 \leq a_k, a_k^* \leq 4$ *for* $k \geq 1$.

Proof. When $N = 4$, we have by Theorem 1 that $CF(4) + CF(4) = [\sqrt{2} - 1, 4(\sqrt{2} - 1)]$, and this interval has length greater than 1.

§5. A theorem of Hurwitz.

In Chapter II, we saw that the solutions of the approximation $b \| bt \| < K$ are given by the convergents of t when K is a sufficiently small positive constant. For $K = 1$, we found that the Ostrowski representation of b consists of no more than two consecutive terms, for $K = 1/2$, it must be a multiple of a B_k, and for $K < 1/2$, we questioned the existence of any solutions at all. In this section, we shall discuss the existence of such solutions for all irrational numbers t, while in Chapter VI we shall continue the discussion for "most" numbers in a sense to be described in Chapter V.

We rewrite $b \| bt \| < K$ as $|t - a/b| < K/b^2$ and consider solutions with $(a, b) = 1$. We begin by taking $K = 1/2$.

Theorem 1. *At least one of every two convergents of t satisfies* $\left| t - \dfrac{a}{b} \right| < \dfrac{1}{2b^2}$.

Proof. We know by (I.2.6) that

$$\left| t - \frac{A_k}{B_k} \right| = \frac{1}{B_k^2(\zeta_{k+1} + \xi_k)}$$

where we use the notation $\xi_k = B_{k-1}/B_k$ of (I.6.2). Thus it suffices to show that it can not happen that $\zeta_{k+1} + \xi_k < 2$ and $\zeta_{k+2} + \xi_{k+1} < 2$. By way of contradiction, suppose that both inequalities are satisfied for some k. From the first, we have by (I.2.4) that

$$a_{k+1} + \frac{1}{\zeta_{k+2}} + \xi_k < 2$$

and so

$$\frac{1}{\zeta_{k+2}} < 2 - (a_{k+1} + \xi_k) = 2 - \frac{1}{\xi_{k+1}},$$

while from the second we have

$$\zeta_{k+2} < 2 - \xi_{k+1}.$$

So

$$1 < \left(2 - \frac{1}{\xi_{k+1}}\right)\left(2 - \xi_{k+1}\right)$$

and hence

$$0 < 2 - \xi_{k+1} - \frac{1}{\xi_{k+1}}.$$

But then $(\xi_{k+1} - 1)^2 < 0$, and this contradiction establishes the result.

Using the same method, we now show a theorem of Hurwitz [1891].

Theorem 2. *At least one of every three consecutive convergents of t satisfies*

$$\left| t - \frac{a}{b} \right| < \frac{1}{\sqrt{5}\, b^2}.$$

Proof. By way of contradiction, let us suppose that we have three consecutive failures:

$$\zeta_{k+1} + \xi_k < \sqrt{5}$$
$$\zeta_{k+2} + \xi_{k+1} < \sqrt{5}$$
$$\zeta_{k+3} + \xi_{k+2} < \sqrt{5}.$$

As in the previous proof, the first two inequalities may be combined to give

$$1 < \left(\sqrt{5} - \frac{1}{\xi_{k+1}} \right)\left(\sqrt{5} - \xi_{k+1} \right)$$

so that

$$\xi_{k+1}^2 - \sqrt{5}\, \xi_{k+1} + 1 < 0.$$

Repeating the same argument with the second and third inequalities, we also have

$$\xi_{k+2}^2 - \sqrt{5}\, \xi_{k+2} + 1 < 0.$$

Thus ξ_{k+1} and ξ_{k+2} are bounded below by $(\sqrt{5}-1)/2$ and above by $(\sqrt{5}+1)/2$. Since $\xi_{k+2} = 1/(a_{k+2} + \xi_{k+1})$ and $a_{k+2} \geq 1$,

$$\xi_{k+2} < \frac{1}{1 + (\sqrt{5}-1)/2} = \frac{\sqrt{5}-1}{2}$$

and the contradiction is obtained.

It now seems natural to expect that for some smaller K, four consecutive terms would need to be considered, and so on, for even smaller values of K. However, Hurwitz also showed that the constant $1/\sqrt{5}$ can not be reduced further by showing that g (the "golden ratio" described in §7 of Chapter I) satisfies any stronger approximation only a finite number of times.

Theorem 3. *Let $g = (1 + \sqrt{5})/2$, as usual. Then*

$$\left| g - \frac{a}{b} \right| < \frac{K}{b^2}$$

has at most a finite number of solutions for any $K < 1/\sqrt{5}$.

Proof. Since $g = [1; 1, 1, \ldots]$ has $g = \zeta_k$ for all $k \geq 0$, we see that $B_{k+1}/B_k = [1; 1, \ldots, 1] \to g$ as $k \to \infty$ and $\xi_k = \frac{1}{g} + \epsilon_k$, where $\epsilon_k \to 0$ as $k \to \infty$. So $\zeta_{k+1} + \xi_k = \sqrt{5} + \epsilon_k$ and

$$\left| g - \frac{A_k}{B_k} \right| = \frac{1}{B_k^2(\sqrt{5} + \epsilon_k)}.$$

Since any $K < 1/\sqrt{5}$ has $K < 1/(\sqrt{5} + \epsilon_k)$ for k sufficiently large, our proof is complete.

Thus if $t \sim g$, then $b \parallel bt \parallel < K$ has at most a finite number of solutions for any $K < 1/\sqrt{5}$, and we may expect a better possible degree of approximation for numbers having infinitely many partial quotients greater than one.

Theorem 4. *Let n be a positive integer. If the three consecutive convergents A_k/B_k, A_{k+1}/B_{k+1}, A_{k+2}/B_{k+2} of t fail to satisfy*

$$\left| t - \frac{a}{b} \right| < \frac{1}{\sqrt{n^2 + 4}\, b^2}$$

then $a_{k+2} < n$.

Proof. As in the proof of Theorem 2, the three consecutive failures mean that

$$\zeta_{k+1} + \xi_k < \sqrt{n^2 + 4}$$
$$\zeta_{k+2} + \xi_{k+1} < \sqrt{n^2 + 4}$$
$$\zeta_{k+3} + \xi_{k+2} < \sqrt{n^2 + 4}$$

and the first two inequalities may be combined to give

$$1 < \left(\sqrt{n^2+4} - \frac{1}{\xi_{k+1}}\right)\left(\sqrt{n^2+4} - \xi_{k+1}\right).$$

Thus

$$\xi_{k+1}^2 - \sqrt{n^2+4}\,\xi_{k+1} + 1 < 0$$

and

$$1 - \sqrt{n^2+4}\,\frac{1}{\xi_{k+1}} + \frac{1}{\xi_{k+1}^2} < 0.$$

Repeating the same argument with the second and third inequalities, we have the same results for ξ_{k+2}. Hence

$$\frac{-n+\sqrt{n^2+4}}{2} < \xi_{k+1},\,\frac{1}{\xi_{k+1}},\,\xi_{k+2},\,\frac{1}{\xi_{k+2}} < \frac{n+\sqrt{n^2+4}}{2}$$

and so

$$\frac{1}{\xi_{k+2}} - \xi_{k+1} < \frac{n+\sqrt{n^2+4}}{2} - \frac{-n+\sqrt{n^2+4}}{2} = n.$$

But this is the same as

$$\frac{1}{B_{k+1}}\,(B_{k+2} - B_k) < n$$

and so $a_{k+2} < n$ as claimed.

Since it is impossible that $a_{k+2} < 1$, the case $n=1$ is merely the contrapositive of Hurwitz's theorem. More generally, we immediately have:

Corollary. *For any irrational number t either there are infinitely many positive integers b such that*

$$b\,\|\,bt\,\| < \frac{1}{\sqrt{n^2+4}}$$

or $a_k < n$ for all sufficiently large indices k.

Let us define $\nu(t)$ for an irrational number t by

$$\nu(t) = \liminf_{b \to \infty}\, b\,\|\,bt\,\|.$$

We have shown that $\nu(t) \le 1/\sqrt{5}$ for any irrational number t, $\nu(t) \le 1/\sqrt{8}$ unless

$t \sim g = [1; 1, 1, \dots]$, $\nu(t) \leq 1/\sqrt{13}$ unless $t \sim [a_0; a_1, a_2, \dots]$ where $a_k < 3$ for all k's, and so on. If $a_k < 3$ for all indices k, then

$$\limsup_{k \to \infty} \left(\zeta_{k+1} + \xi_k \right) \leq [2; \overline{1,2}\,] + [0; \overline{1,2}\,] = \sqrt{12}$$

and so if $\nu(t) > 1/\sqrt{13}$, it must be $\geq 1/\sqrt{12}$ and it is impossible for $\nu(t)$ to take any value between these two quantities. In the next section, we shall undertake a detailed study of the values of $\nu(t)$.

§6. The Lagrange and Markov spectra.

We saw in the previous section that for any irrational t the inequality

$$b \, \| \, bt \, \| \; < \; \frac{1}{\sqrt{5}} \tag{1}$$

has infinitely many integer solutions and if t is equivalent to a root of the quadratic equation

$$x^2 + x - 1 \; = \; 0,$$

then the constant $1/\sqrt{5}$ can not be decreased. If t is not equivalent to such a root, then the inequality

$$b \, \| \, bt \, \| \; < \; \frac{1}{\sqrt{8}} \tag{2}$$

has infinitely many integer solutions and if t is equivalent to a root of the quadratic equation

$$x^2 + 2x - 1 \; = \; 0,$$

then the constant $1/\sqrt{8}$ can not be decreased. By the results in Chapter III on periodic continued fractions and quadratic surds, it is natural to expect that this close correspondence between the possible degree of approximation in the inequality $b \, \| \, bt \, \| \, < 1/K$ and quadratic surds may be extended further. In the last section, we introduced the quantity

$$\nu(t) \; = \; \liminf_{b \to \infty} \; b \, \| \, bt \, \|$$

and studied some values of ν by calculating

$$\limsup_{k\to\infty} \left(\zeta_{k+1} + \xi_k\right)$$

for the continued fractions of several values of t. The *Lagrange spectrum* $\mathfrak{L}$ is the set of all possible values of ν.

Closely connected with the Lagrange spectrum are the values of the quadratic forms corresponding to the quadratic expressions $x^2 + x - 1$ and $x^2 + 2x - 1$ mentioned above. A *real, indefinite binary quadratic form* is an expression

$$f(x,y) = \alpha x^2 + \beta xy + \gamma y^2$$

in two variables x and y with real coefficients α, β and γ such that the *discriminant* $d = d(f) = \beta^2 - 4\alpha\gamma$ is positive. Thus $f(x,y)$ has two distinct roots

$$\theta = \frac{-\beta + \sqrt{d}}{2\alpha} \quad \text{and} \quad \varphi = \frac{-\beta - \sqrt{d}}{2\alpha}, \tag{3}$$

and factors as $f(x,y) = \alpha(x - \theta y)(x - \varphi y)$. We shall assume that θ and φ are irrational. Two forms f and f^* are *equivalent*, $f \sim f^*$, if

$$f^*(x,y) = f(Ax + By, Cx + Dy),$$

where A, B, C and D are integers such that $AD - BC = \pm 1$. Thus $f^*(x,y) = \alpha^* x^2 + \beta^* xy + \gamma^* y^2$ is really

$$f^*(x,y) = f(A,C)x^2 + \left(2AB\alpha + (AD + BC)\beta + 2CD\gamma\right)xy + f(B,D)y^2 \tag{4}$$

and the discriminant of f^* is the same as that of f. Further, we also have $f(x,y) = f^*(Dx - By, -Cx + Ay)$ and, since $f \sim f^*$ and $f^* \sim f^{**}$ yield $f \sim f^{**}$ by the usual composition of unimodular transformations, this equivalence of forms is an equivalence relation.

Let

$$\mu(f) = \inf_{(x,y) \neq (0,0)} |f(x,y)|,$$

where x and y are integers. If $f \sim f^*$, then $\mu(f) = \mu(f^*)$. Since $\mu(\lambda f) = \lambda \mu(f)$ and $d(\lambda f) = \lambda^2 d(f)$ for any non-zero real number λ, the value of

$$\frac{\mu(f)}{\sqrt{d(f)}}$$

is the same for equivalent forms and non-zero multiples of them. The *Markov spectrum* $\mathfrak{M}$ is the set of all values of $\mu(f)/\sqrt{d(f)}$. From this viewpoint, (1) may be restated as

$$\frac{\mu(f)}{\sqrt{d(f)}} \leq \frac{1}{\sqrt{5}}$$

for any form f with equality only if f is equivalent to a multiple of $x^2 + xy - y^2$, and for forms not of this type, (2) becomes

$$\frac{\mu(f)}{\sqrt{d(f)}} \leq \frac{1}{\sqrt{8}}$$

with equality only if f is equivalent to a multiple of $x^2 + 2xy - y^2$.

In this section, we will develop some basic properties of indefinite binary quadratic forms, reformulate the definition of the Markov spectrum in terms of continued fractions, and deduce some important features of these spectra.

From the factorization of $f(x, y)$ as $\alpha(x - \theta y)(x - \varphi y)$, any equivalent form $f^*(x, y)$ has the factorization

$$\alpha\Big((Ax + By) - \theta(Cx + Dy)\Big)\Big((Ax + By) - \varphi(Cx + Dy)\Big)$$

$$= \alpha(A - C\theta)(A - C\varphi)\left(x - \frac{-B + D\theta}{A - C\theta}y\right)\left(x - \frac{-B + D\varphi}{A - C\varphi}y\right).$$

Since

$$\frac{-B + \dfrac{-\beta \pm \sqrt{d}}{2\alpha}D}{A - \dfrac{-\beta \pm \sqrt{d}}{2\alpha}C} = \frac{-\beta^* \pm (AD - BC)\sqrt{d}}{\alpha^*},$$

the roots θ^* and φ^* of f^* are equivalent in the sense of §2 of Chapter I to θ and φ if $AD - BC = 1$ and to φ and θ if $AD - BC = -1$. We have the pairs of formulas

$$\theta^* = \frac{-B + D\theta}{A - C\theta} \quad \text{and} \quad \varphi^* = \frac{-B + D\varphi}{A - C\varphi}$$

$$\theta = \frac{A\theta^* + B}{C\theta^* + D} \quad \text{and} \quad \varphi = \frac{A\varphi^* + B}{C\varphi^* + D} \tag{5}$$

A form $f(x,y)$ *properly represents* the real number λ if $f(X,Y) = \lambda$ for relatively prime integers X and Y.

Lemma 1. *If f represents λ properly, then f is equivalent to a form f^* with $\alpha^* = \lambda$.*

Proof. Since we may take the integer Y to be positive, the rational number X/Y can be expressed as a continued fraction $[a_0; a_1, a_2, \ldots, a_k]$ with k odd, so that $A_k B_{k-1} - A_{k-1} B_k = 1$. Then $f \sim f^*$ where

$$f^*(x,y) = f(A_k x + A_{k-1} y, B_k x + B_{k-1} y) = \lambda x^2 + \beta^* xy + \gamma^* y^2,$$

since $\alpha^* = f(A_k, B_k) = f(X,Y)$.

Thus the study of $\mu(f)$ is equivalent to the examination of the possible values of α^* for forms f^* equivalent to the given form f. The form $f(x,y) = \alpha(x - \theta y)(x - \varphi y)$ is *reduced* if $|\theta| < 1$, $|\varphi| > 1$ and $\theta\varphi < 0$.

Lemma 2. *f is reduced if and only if $0 < -\beta + \sqrt{d} < 2|\alpha| < \beta + \sqrt{d}$.*

Proof. Suppose first that f is reduced. Since $\theta\varphi = \gamma/\alpha$, we have that α and γ are of opposite signs and $|\beta| < \sqrt{d}$. Further, $\theta\varphi = (-\beta + \sqrt{d})(-\beta - \sqrt{d})/(4\alpha^2) < 0$ means $-\beta + \sqrt{d}$ and $\beta + \sqrt{d}$ are of the same sign. Since $|\theta| < 1$, $|-\beta + \sqrt{d}| < 2|\alpha|$, while $|\varphi| > 1$ means $2|\alpha| < |\beta + \sqrt{d}|$. Then $|-\beta + \sqrt{d}| < |\beta + \sqrt{d}|$ together with $|\beta| < \sqrt{d}$ forces $\beta > 0$, and the inequality follows. Conversely, given the inequality, it must be that β is both positive and less than $\sqrt{d}$, which means $\alpha\gamma < 0$ and so f is reduced.

We also see that if f is reduced, then θ and α have the same sign and γ has the opposite sign. Since $4\alpha\gamma = \beta^2 - d < 0$, the lemma also holds for the inequality $0 < -\beta + \sqrt{d} < 2|\gamma| < \beta + \sqrt{d}$.

Lemma 3. *Every form is equivalent to a form* $f(x,y) = \alpha x^2 + \beta xy + \gamma y^2$ *such that* $|\beta| \le |\alpha| \le \sqrt{d/3}$.

Proof. If $|\alpha| > \sqrt{d/3}$, we consider $f^*(x,y) = f(nx+y, -x)$ where n is an integer. By (4),

$$f^*(x,y) = \alpha^* x^2 + (2\alpha n - \beta)xy + \alpha y^2$$

and we may choose n such that $\beta^* = 2\alpha n - \beta$ satisfies $|\beta^*| \le |\alpha|$. Since the discriminants of f^* and f are the same, $4\alpha^*\alpha = \beta^{*2} - d < \beta^{*2} \le \alpha^2$, $-4\alpha^*\alpha = d - \beta^{*2} < d$, and we have supposed that $d < 3\alpha^2$. Thus $4|\alpha^*\alpha| < 3\alpha^2$ and so $|\alpha^*| < (3/4)|\alpha|$. Repeated application of this transformation can reduce the size of $|\alpha^*|$ below any given positive bound and thus an equivalent form has been found with $|\alpha| \le \sqrt{d/3}$.

Suppose now that the form $f(x,y)$ has $|\alpha| \le \sqrt{d/3}$ but not $|\beta| \le |\alpha|$. Then $f^*(x,y) = f(x+ny, y)$ is, again by (4),

$$f^*(x,y) = \alpha x^2 + (2\alpha n + \beta)xy + \gamma^* y^2$$

and we may choose the integer n such that $\beta^* = 2\alpha n + \beta$ has $|\beta^*| \le |\alpha|$.

Lemma 4. *Every form is equivalent to a reduced form.*

Proof. From Lemma 3, we may suppose that $f(x,y) = \alpha x^2 + \beta xy + \gamma y^2$ has $|\beta| \le \sqrt{d}$. Since θ and φ are not rational, $\alpha\gamma \ne 0$ and $4|\alpha\gamma| = d - \beta^2 < d$ implies that at least one of $2|\alpha|$ and $2|\gamma|$ is $< \sqrt{d}$. We may take $2|\alpha| < \sqrt{d}$ because $f(x,y) \sim f(y,x)$. Since $0 < \sqrt{d} - 2|\alpha| < \sqrt{d}$, there is an integer n with $\sqrt{d} - 2|\alpha| < \beta - 2\alpha n < \sqrt{d}$. We let $f^*(x,y) = f(x-ny, y)$ so, by (4) again,

$$f^*(x,y) = \alpha x^2 + (\beta - 2\alpha n)xy + \gamma^* y^2$$

and $\beta^* = \beta - 2\alpha n$ satisfies $0 < -\beta + \sqrt{d} < 2\,|\,\alpha\,| < \beta + \sqrt{d}$. By Lemma 2, f^* is reduced.

A *right neighbor* of $f(x,y) = \alpha x^2 + \beta xy + \gamma y^2$ is an equivalent form f_{right} such that

$$f_{right}(x,y) \;=\; f(y, -x + ny)$$

for some integer n. By (4),

$$f_{right}(x,y) \;=\; \gamma x^2 + \beta^* xy + \gamma^* y^2$$

with $\beta^* = -(\beta + 2\gamma n)$. Similarly, a *left neighbor* of $f(x,y)$ is a form f_{left} such that

$$f_{left}(x,y) \;=\; f(nx - y, x) \;=\; \alpha^* x^2 + \beta^* xy + \alpha y^2$$

for some integer n, and we have $\beta^* = -(\beta + 2\alpha n)$. Since we are interested in the possible values of the first coefficients of forms equivalent to f, these right and left neighbors will be useful in organizing the equivalent forms for study.

Lemma 5. *Every reduced form has unique reduced right and left neighbors.*

Proof. Since $f(x,y) \sim f(y,x)$, it suffices to show the existence of a unique reduced right neighbor of a given reduced form. Let θ and φ be the roots (3) of the reduced form $f(x,y) = \alpha x^2 + \beta xy + \gamma y^2$. Then θ and α have the same sign and γ has the opposite sign. Let n be the integer with the same sign as θ and such that $|\,n\,| = [1/|\,\theta\,|]$. Since $|\,\theta\,| < 1$, n is not zero and thus

$$f_{right}(x,y) \;=\; f(y, -x + ny)$$

has first root

$$\theta_{right} \;=\; n - \frac{1}{\theta},$$

by (5). Therefore $|\,\theta_{right}\,| < 1$ and the sign of θ_{right} is opposite that of n and θ. Since the sign of φ is the opposite of the sign of θ and n, the second root of f_{right} is

$$\varphi_{right} = n - \frac{1}{\varphi}$$

and has the same sign as n. Hence $|\varphi_{right}| > 1$. Thus f_{right} is reduced, and we have found a reduced right neighbor of f.

We now must show that this reduced form f_{right} is unique. Since f and f_{right} are both to be reduced, θ_{right} must have the same sign as γ and θ must have sign opposite to that of γ. Then $|\theta| < 1$, $|\theta_{right}| < 1$ and (5) imply that n has the same sign as θ and that $|n|$ is the largest integer less than $1/|\theta|$. Thus n, and hence f_{right}, are uniquely determined.

Given any reduced form f, we see from Lemma 5 that it is an element of a *chain* of equivalent, reduced forms

$$\ldots f_{-2}, f_{-1}, f_0, f_1, f_2, \ldots,$$

where f_{k-1} and f_{k+1} are the unique left and right reduced neighbors of f_k. Since the first coefficients of the forms in the chain alternate in sign, we may choose $f_0(x,y)$ so that it has a positive first coefficient and then the forms in the chain are

$$f_k(x,y) = (-1)^k \alpha_k x^2 + \beta_k xy + (-1)^{k+1} \alpha_{k+1} y^2$$

with

$$f_{k+1}(x,y) = f_k(y, -x + n_k y)$$

so that α_k, β_k and $(-1)^k n_k$ are positive for every k. Let b_k be the positive integer $(-1)^k n_k$, let

$$\theta_k = \frac{-\beta_k + \sqrt{d}}{2(-1)^k \alpha_k} \quad \text{and} \quad \varphi_k = \frac{-\beta_k - \sqrt{d}}{2(-1)^k \alpha_k}$$

be, as in (3), the roots of $f_k(x,y)$, and let

$$R_k = \frac{(-1)^k}{\theta_k} \quad \text{and} \quad S_k = \frac{(-1)^{k+1}}{\varphi_k}$$

so that

$$R_k = \frac{\beta_k + \sqrt{d}}{2\alpha_{k+1}} \quad \text{and} \quad S_k = \frac{-\beta_k + \sqrt{d}}{2\alpha_{k+1}}.$$

Thus $R_k > 1$ and $0 < S_k < 1$. By (5), we have

$$\frac{1}{\theta_k} = n_k - \theta_{k+1} \quad \text{and} \quad \frac{1}{\varphi_k} = n_k - \varphi_{k+1},$$

which means that

$$R_k = b_k + \frac{1}{R_{k+1}} \quad \text{and} \quad S_k + b_k = \frac{1}{S_{k+1}}.$$

Thus R_k and S_k are the continued fractions

$$R_k = [b_k; b_{k+1}, b_{k+2}, \dots] \quad \text{and} \quad S_k = [0; b_{k-1}, b_{k-2}, \dots]$$

and

$$R_k + S_k = \frac{\sqrt{d}}{\alpha_{k+1}}. \tag{6}$$

The roots θ_0 and φ_0 of f_0 are thus connected to the roots of every f_k in the chain by the continued fractions

$$R_0 = [b_0; b_1, \dots, b_{k-1}, R_k] \tag{7}$$

and

$$S_k = [0; b_{k-1}, b_{k-2}, \dots, b_0 + S_0]. \tag{8}$$

We may expect that these continued fractions will play the same role for the Markov spectrum that ζ_{k+1} and ξ_k played for the Lagrange spectrum. First, however, we need:

Lemma 6. *Any two equivalent reduced forms belong to the same chain.*

Proof. Since either form may be replaced by its right reduced neighbor, it suffices to show the claim for two distinct reduced forms $f(x,y) = \alpha x^2 + \beta xy + \gamma y^2$ and $f^*(x,y) = \alpha^* x^2 + \beta^* xy + \gamma^* y^2$ that are equivalent and have positive first coefficients; that is, $\alpha, \alpha^* > 0$. Thus θ and θ^* are $0 < \theta$, $\theta^* < 1$, φ and φ^* are φ, $\varphi^* < -1$, and $f^*(x,y) = f(Ax + By, Cx + Dy)$ with integers A, B, C and D such that

$AD - BC = 1$. By changing the signs of A, B, C and D if necessary, we may suppose that either $A = 0$ and $C \geq 1$, or $A > 0$.

If $A = 0$ and $C \geq 1$, then $-BC = 1$ means that $B = -1$ and $C = 1$. Then by (5),

$$\theta \; = \; \frac{-1}{\theta^* + D} \quad \text{so} \quad -D \; = \; \tfrac{1}{\theta} + \theta^* > 1,$$

while

$$\varphi \; = \; \frac{-1}{\varphi^* + D} \quad \text{gives} \quad D \; = \; \frac{-1}{\varphi} - \varphi^* > 1,$$

which is a contradiction. Thus $A \neq 0$ and we may suppose $A > 0$. Again by (5),

$$\left(\tfrac{A}{\theta} - C\right)\!\left(A\theta^* + B\right) \; = \; 1 \tag{9}$$

and, similarly,

$$\left(\tfrac{A}{\varphi} - C\right)\!\left(A\varphi^* + B\right) \; = \; 1. \tag{10}$$

If $C = 0$, then $A = D = 1$ and (9) becomes $(\theta^* + B)/\theta = 1$. Thus $|B| < 1$, because $B = \theta - \theta^*$, and so $B = 0$. If $B = 0$, then again $A = D = 1$ and (10) becomes $(1/\varphi - C)\varphi^* = 1$; thus $|C| < 1$, because $-C = (1/\varphi^*) - (1/\varphi) < (-1/\varphi) < 1$ and $C = (-1/\varphi^*) + (1/\varphi) < (-1/\varphi^*) < 1$, and so $C = 0$. In either case, we obtain $f^*(x,y) = f(x,y)$, contrary to the assumption that f and f^* were distinct. Thus B and C are both non-zero, and we have established that $A > 0$ and $BC \neq 0$.

If $B > 0$, then $A\theta^* + B > 1$ and thus $0 < (A/\theta) - C < 1$ by (9). Then $A < (A/\theta) < 1 + C$ gives $C > 0$. If $C > 0$, then $(A/\varphi) - C < -1$ and so $-1 < A\varphi^* + B < 0$ by (10). Then $A < A(-\varphi^*) < 1 + B$ gives $B > 0$. Hence not only is $BC \neq 0$, but $BC > 0$. Since $AD = 1 + BC$ and $A > 0$, we see that $D > 0$. We now may suppose that A, B, C and D are all positive integers, since if B and C were negative, we could look instead at the "inverse" equivalence $f(x,y) = f^*(Dx - By, -Cx + Ay)$.

We now have that the positive integers A, B, C and D satisfy $AD - BC = 1$, $A < C$ and $A < B$. Since $AD > BC$, we also have that $C < D$ and $B < D$. The integers D and B are relatively prime since $AD - BC = 1$, and the continued fraction

$$\frac{D}{B} \; = \; [a_0; a_1, \dots, a_k]$$

has $A_k = D$ and $B_k = B$. Since $a_k = 1$ or not as we choose, we may take k odd so that $A_k B_{k-1} - A_{k-1} B_k = 1$. Since the relatively prime positive integers A and C satisfy $AD - BC = 1$ and are such that $A < B$ and $C < D$, we see that $A = B_{k-1}$ and $C = A_{k-1}$. But then (again using (5)),

$$\frac{1}{\theta} = \frac{D\left(\frac{1}{\theta^*}\right) + C}{B\left(\frac{1}{\theta^*}\right) + A}$$

means that $1/\theta = [a_0; a_1, \dots, a_k, 1/\theta^*]$ and, by the uniqueness of the continued fraction (7) of $1/\theta$, θ^* must be the first root of a form f^{**} that is $k + 1$ terms to the right of f in its chain. Since

$$\frac{D}{C} = \frac{A_k}{A_{k-1}} = [a_k; a_{k-1}, \dots, a_0] \tag{11}$$

and

$$\frac{B}{A} = \frac{B_k}{B_{k-1}} = [a_k; a_{k-1}, \dots, a_1],$$

by §6 of Chapter I, the continued fraction (8) shows that the second root of f^{**} is the same as φ^*. Thus f^* and f^{**} have the same roots and discriminant and hence $f^{**} = f^*$.

We can now show the following theorem of Lagrange.

Theorem 1. *Let* $\dots, f_{-1}, f_0, f_1, \dots$ *be a chain of reduced forms with* $\alpha_0 > 0$ *and let* λ *be properly represented by a form in the chain. If* $|\lambda| < \sqrt{d}/2$, *then* $\alpha_k = |\lambda|$ *for some* k.

Proof. Let $|\lambda| < \sqrt{d}/2$ be properly represented by a form in the chain. By Lemma 1, there is an equivalent form $f(x,y) = \lambda x^2 + \beta xy + \gamma y^2$ and, by the proof of Lemma 4, there is an equivalent reduced form $f^*(x,y) = \lambda x^2 + \beta^* xy + \gamma^* y^2$, which is a member of the chain by Lemma 6.

Corollary 1. *Let* $\dots, f_{-1}, f_0, f_1, \dots$ *be a chain of reduced forms with* $\alpha_0 > 0$ *and let* f *be a form equivalent to* f_0. *Then* $\mu(f) = \inf_k \alpha_k$.

Proof. Since $f(x,y)$ is equivalent by Lemma 4 to a reduced form $\alpha x^2 + \beta xy + \gamma y^2$

with $\alpha\gamma < 0$, the smaller of $|\alpha|$ and $|\gamma|$ is $< \sqrt{d}/2$. Thus f properly represents a number λ with $|\lambda| < \sqrt{d}/2$ and the result follows.

Since each α_{k+1} from the chain is

$$\frac{\alpha_{k+1}}{\sqrt{d}} = \frac{1}{R_k + S_k}$$

by (6), and each of the numbers in the Lagrange spectrum is of the form

$$\frac{1}{\zeta_{k+1} + \xi_k},$$

we also have shown:

Corollary 2. *Let $\mathfrak{B}$ be the set of all doubly infinite sequences*

$$b = \{ \ldots, b_{-1}, b_0, b_1, \ldots \}$$

of positive integers b_k and let

$$\lambda_k(b) = [b_k; b_{k+1}, \ldots] + [0; b_{k-1}, b_{k-2}, \ldots].$$

Then the Markov spectrum $\mathfrak{M}$ is $\mathfrak{M} = \left\{ \inf_k \dfrac{1}{\lambda_k(b)} \right\}_{b \in \mathfrak{B}}$ *and the Lagrange spectrum $\mathfrak{L}$ is* $\mathfrak{L} = \left\{ \liminf_{k \to \infty} \dfrac{1}{\lambda_k(b)} \right\}_{b \in \mathfrak{B}}.$

With this alternative formulation of $\mathfrak{M}$, we may now show:

Theorem 2. *The largest values of $\mathfrak{M}$ are $1/\sqrt{5}$ and $1/\sqrt{8}$, $\mathfrak{M}$ contains no value in the interval $(1/\sqrt{13}, 1/\sqrt{12})$, and $\mathfrak{M}$ contains every value in the interval $(0, 4/(10 + 9\sqrt{2}))$.*

Proof. The largest values of $\mathfrak{M}$ and the existence of a gap between $1/\sqrt{13}$ and $1/\sqrt{12}$ follow immediately from Theorem 5.2 and the discussion of $\nu(t)$ after Theorem 5.4

and its Corollary. That $\mathfrak{M}$ contains the interval $(0, 4/(10 + 9\sqrt{2}))$ follows from Hall's theorem (the Corollary to Theorem 4.1), since if $1/\lambda < 4/(10 + 9\sqrt{2})$ then λ is greater than

$$(10 + 9\sqrt{2})/4 \; = \; 4 + [0; \overline{1,4}\,] + [0; 1, 5, \overline{1,4}\,]$$

and thus can be written in the form

$$\lambda \; = \; n + [0; a_1, a_2, \ldots\,] + [0; a_1^*, a_2^*, \ldots\,],$$

where $n \geq 5$ and the partial quotients satisfy $1 \leq a_k, \; a_k^* \leq 4$ for $k \geq 1$. Then the number

$$\lambda^* \; = \; [n; \; a_1^*, n, a_1, \; a_2^*, a_1^*, n, a_1, a_2, \; a_3^*, a_2^*, a_1^*, n, a_1, a_2, a_3, \; \ldots\,]$$

has

$$\limsup_{k \to \infty} \zeta_{k+1} + \xi_k \; = \; \lim_{k \to \infty} \zeta_{k(k+1)} + \xi_{k(k+1)-1}$$

$$= \; \lim_{k \to \infty} n + [0; a_1, a_2, \ldots\,] + [0; a_1^*, a_2^*, \ldots\,] \; = \; \lambda,$$

as desired.

The part of the Markov spectrum above $1/\sqrt{12}$ corresponds to doubly infinite sequences $\{ \ldots, b_{-1}, b_0, b_1, \ldots \}$ consisting only of 1's and 2's. We shall write such sequences as strings of 1's and 2's and, to retain the form of the $\zeta_{k+1} + \xi_k$ notation for $\nu(t)$, we will indicate the position of the partial quotient "a_{k+1}" with an arrow; thus the value

$$\cdots 211112\underset{\uparrow}{2}11112211 \cdots$$

is to be interpreted as $[2; 1, 1, 1, 1, 2, 2, 1, 1, \ldots\,] + [0; 2, 1, 1, 1, 1, 2, \ldots\,]$.

Theorem 3. *There are uncountably many non-equivalent numbers t such that* $\nu(t) = 1/3$.

Proof. First, consider the number

$$t = [2; 2, \; 1, \; 2, 2, \; 1, 1, \; 2, 2, \; 1, 1, 1, \; 2, 2, \; 1, 1, 1, 1, \; 2, 2, \; \ldots\,],$$

where the partial quotients have each pair of 2's separated from the next by one more 1 than before. To calculate $\nu(t)$, we need only consider $\zeta_{k+1} + \xi_k$ starting at a 2 and then

$$\cdots 111\underset{\uparrow}{2}21111 \cdots$$

gives

$$\zeta_{k+1} + \xi_k = [2; 1, 1, 1, 1, \ldots] + [0; 2, 1, 1, 1, \ldots]$$
$$\to 2 + \frac{1}{g} + \frac{1}{2 + \frac{1}{g}} = 3,$$

while

$$\cdots 111\underset{\uparrow}{2}21111 \cdots$$

gives

$$\zeta_{k+1} + \xi_k = [2; 2, 1, 1, 1, 1, \ldots] + [0; 1, 1, 1, \ldots]$$
$$\to 2 + \frac{1}{2 + \frac{1}{g}} + \frac{1}{g} = 3.$$

Any other number of the form

$$[2;2, \underbrace{1,\ldots,1}_{r_1}, 2,2, \underbrace{1,\ldots,1}_{r_2}, 2,2, \underbrace{1,\ldots,1}_{r_3}, 2,2, \underbrace{1,\ldots,1}_{r_4}, 2,2, 1,\ldots],$$

where the blocks of 1's between the pairs of 2's have increasing lengths r_k, will also have $\nu = 1/3$. Furthermore, the sequences $\{r_k\}$ and $\{r_k'\}$ for two such numbers will be identical from some point onward if and only if the numbers are equivalent. Thus the notion of equivalence of continued fractions naturally extends to equivalence between sequences $\{r_k\}$.

Let $\Re$ be the set of all non-equivalent sequences $\{r_k\}$ that are strictly increasing. Thus $\Re$ contains, for example, the sequence $R_1 = \{1, 2, 3, \ldots\}$ of all positive integers, which corresponds to the number t above. To show the theorem, it suffices to show that $\Re$ is uncountable; so, by way of contradiction, let us suppose that $\Re$ is countable. Then the sequences of $\Re$ may be listed as R_1, R_2, R_3, $\ldots$ and each R_k for $k > 1$ omits infinitely many positive integers, since it is not equivalent to R_1. Let $S = \{s_k\}$ be the increasing sequence of positive integers with $s_1 = 1$ and s_{k+1}, for each $k > 0$, being the smallest integer not in $R_2, \ldots, R_{k+1}$ such that $s_{k+1} > s_k + 1$. Since S is not equivalent to R_1 and contains infinitely many positive integers not in R_k for each $k > 1$, it is not equivalent to any of the sequences in the list, thus contradicting the assumption that $\Re$ was countable.

We now turn to the part of $\mathfrak{M}$ and $\mathfrak{L}$ greater than 1/3. We have already described the largest values, $\nu([\bar{1}]) = 1/\sqrt{5}$ and $\nu([\bar{2}]) = 1/\sqrt{8}$, so let us consider sequences containing both 1's and 2's such that $\nu > 1/3$. If we have

$$\cdots 1\underset{\uparrow}{2}1 \cdots ,$$

then

$$\zeta_{k+1} + \xi_k = [2; 1, \ldots] + [0; 1, \ldots]$$
$$> [2; 2] + [0; 2] = 3,$$

while if we have $\cdots 212 \cdots$, then it can not be $\cdots 1212 \cdots$ and thus must be

$$\cdots 2\underset{\uparrow}{2}12 \cdots .$$

But then

$$\zeta_{k+1} + \xi_k = [2; 1, 2, \ldots] + [0; 2, \ldots]$$
$$> [2; 1, 2] + [0; 2, 1] = 3.$$

Consequently, if t is not equivalent to either $[\bar{1}]$ or $[\bar{2}]$, then the sequence must contain $\cdots 1122 \cdots$ infinitely many times. From the proof of Theorem 3, we see that these "1122" patterns may not be separated by blocks of 1's with unbounded lengths. Moreover, they may not be followed by blocks of 2's with unbounded lengths, since if they were, we would have from

$$\cdots 1\, 1\, \underset{\uparrow}{2}\, 2\, \underbrace{2 \cdots 2} \cdots$$

that

$$\zeta_{k+1} + \xi_k = [2; 2, \underbrace{2, \ldots, 2}, \ldots] + [0; 1, 1, \ldots]$$
$$> [2; 2, \underbrace{2, \ldots, 2}, \ldots] + [0; 1, 1, 2] \rightarrow (1 + \sqrt{2}) + \frac{3}{5} > 3.$$

Thus the 1122 patterns are separated by blocks with bounded length.

A simple calculation shows that $[2; 2, x] + [0; 1, 1, y] < 3$ for real numbers $x > 1$ and $y > 1$ if and only if $x < y$. For a sequence containing 1122 infinitely many times, we must examine $\zeta_{k+1} + \xi_k$ at the first 2 in each of these patterns:

where the x_{k_i} are the subsequent terms of ζ_{k_i+1} starting with the first digit to the right of the 22 and going to the right and the y_{k_i} are the prior terms of ξ_{k_i} starting with the first digit to the left of the 11 and going to the left. Since $\nu > 1/3$, we must have $x_{k_i} < y_{k_i}$. For the sequence to represent the value of lim sup $\zeta_{k+1} + \xi_k$, it must be that x_{k_i} and $y_{k_{i+1}}$ contain the same terms for each i. But since the first difference between x_{k_i} and $y_{k_{i+1}}$ occurs when x_{k_i} has a 1 while $y_{k_{i+1}}$ has a 2, these different numbers must occur after an even number of terms. Thus it suffices to consider the values corresponding to purely periodic continued fractions of the form

$$[2;2,\ a_1, a_2,\ \ldots,\ a_p, a_p,\ \ldots,\ a_2, a_1,\ 1, 1],$$

where $p \geq 0$ is an integer and the partial quotients a_k are no larger than 2. Our next two lemmas depend only on the symmetry of the block between the 22 and the 11 and not on the size of the partial quotients within the block.

Lemma 7. $[2;2,\ a_1, a_2,\ \ldots,\ a_p, a_p,\ \ldots,\ a_2, a_1,\ 1, 1]$ *is a root of an equation of the form*

$$mx^2 + (3m - 2k)x + (l - 3k) = 0,$$

where the positive integers k, l and m satisfy $k^2 + 1 = lm$.

Proof. Let $n + 1$ be the period length of

$$x = [2;2,\ a_1, a_2,\ \ldots,\ a_p, a_p,\ \ldots,\ a_2, a_1,\ 1, 1].$$

Then by (III.1.4), x satisfies

$$B_n x^2 - (A_n - B_{n-1})x - A_{n-1} = 0,$$

where

$$\frac{A_n}{B_n} = [2; 2, a_1, a_2, \ldots, a_p, a_p, \ldots, a_2, a_1, 1, 1].$$

We also have, as in (11), that

$$\frac{A_n}{A_{n-1}} = [1; 1, a_1, a_2, \ldots, a_p, a_p, \ldots, a_2, a_1, 2, 2].$$

Writing $A(r)$ for the "lowest terms" numerator of the rational number r, we see that

$$A([2; 2, a_1, \ldots, a_1, 1, 1]) = A([1; 1, a_1, \ldots, a_1, 2, 2]),$$

which is then

$$= 2A([1; 1, a_1, \ldots, a_1, 2]) + A([1; 1, a_1, \ldots, a_1])$$
$$= 2A([1; 1, a_1, \ldots, a_1, 2]) + A([a_1; a_2, \ldots, a_2, a_1, 1, 1])$$
$$= 2A([2; a_1, \ldots, a_1, 1, 1]) + A([a_1; a_2, \ldots, a_2, a_1, 2]).$$

Since

$$A([2; a_1, \ldots, a_1, 1, 1]) = A([2; a_1, \ldots, a_1, 1]) + A([2; a_1, a_2, \ldots, a_2, a_1]),$$

we obtain

$$A([2; 2, a_1, \ldots, a_1, 1, 1]) = 3A([2; a_1, \ldots, a_1, 1, 1]) - A([2; a_1, \ldots, a_1, 1]),$$

and thus $A_n = 3B_n - B_{n-1}$. Since $A_n B_{n-1} - A_{n-1} B_n = 1$, we may rewrite our quadratic equation in x as

$$B_n x^2 - (A_n - B_{n-1})x - \frac{A_n B_{n-1} - 1}{B_n} = 0,$$

which then becomes

$$B_n x^2 + (3B_n - 2A_n)x + \left(\frac{A_n^2 + 1}{B_n} - 3A_n\right) = 0$$

and is of the required form with $k = A_n$, $l = (A_n^2 + 1)/B_n$ and $m = B_n$.

For example, $[2; 2, 1, 1, 1, 1, 2, 2, 1, 1, 1, 1, 1, 1]$ has $p = 6$ and $n = 13$ with $A_n = 3157$ and $B_n = 1325$. Thus we have a root of the equation

$$1325x^2 - 2339x - 1949 = 0.$$

However, the associated quadratic form

$$f(x, y) = 1325x^2 - 2339xy - 1949y^2$$

is not reduced. Taking $\alpha = 1325$, $\beta = -2339$ and $\gamma = -1949$ in the proof of Lemma 4, we have $d = 15800621$ so that $|\beta|, 2|\alpha| < \sqrt{d}$, and we need the integer n to be such that

$$0 < -2650 + \sqrt{15800621} < -2339 - 2650n < \sqrt{15800621}.$$

Thus $n = -2$ and the form

$$f^*(x, y) = f(x + 2y, y) = 1325x^2 + 2961xy - 1327y^2$$

is reduced. We now show that this choice for n is the correct choice in general.

Lemma 8. $[2; 2, a_1, a_2, \ldots, a_p, a_p, \ldots, a_2, a_1, 1, 1]$ *is a root of* $f(x, 1) = 0$, *where the form* $f(x, y)$ *is equivalent to* $f^*(x, y) = f(x + 2y, y)$ *such that*

$$f^*(x, y) = mx^2 + (3m - 2k)xy + (l - 3k)y^2$$

is reduced and the positive integers k, l *and* m *satisfy* $0 < 2k < m$ *and* $k^2 + 1 = lm$.

Proof. From the proof of Lemma 7, we have that

$$f^*(x, y) = B_n x^2 + \left(3B_n - 2(A_n - 2B_n)\right)xy + \left(\frac{A_n^2 + 1}{B_n} + 4B_n - 4A_n - 3(A_n - 2B_n)\right)y^2.$$

Thus $k = A_n - 2B_n$ satisfies $0 < 2k < m$, because $2 < A_n/B_n < 5/2$, and $l = (A_n^2 + 1)/B_n + 4B_n - 4A_n$ has $lm = A_n^2 + 1 + 4B_n^2 - 4A_nB_n = (A_n - 2B_n)^2 + 1$, as required. That f^* is reduced follows from the requirement given in Lemma 2.

It should be noted that while necessary, our restriction to symmetric patterns $[\overline{2;2},\ a_1, a_2,\ \ldots,\ a_p, a_p,\ \ldots,\ a_2, a_1,\ \overline{1,1}]$ is not sufficient to ensure $\zeta_{k+1} + \xi_k < 3$. For example, $[\overline{2;2},\ 1,\ 1,\ 2,\ 2,\ 2,\ 2,\ 1,\ 1,\ \overline{1,1}]$ corresponds to

$$\cdots 1\ 1\ 1\ 1\ \underset{\underset{\zeta_{k+1}}{\uparrow}}{2}\ 2\ 1\ 1\ 2\ 2\ 2\ \underset{\underset{\zeta_{k+8}}{\uparrow}}{2}\ 1\ 1\ 1\ 2\ 2\ 1\ 1\ 2\ 2\ 2\ 2\ 1\ 1\ 1\ 1\ \cdots$$

and, while $\zeta_{k+1} + \xi_k = \sqrt{2873024}/565 < 3$, the value of $\zeta_{k+8} + \xi_{k+7}$ is $\sqrt{2873024}/560 > 3$. One clue to the difference in values between this example and the previous is that while the symmetric interior of 22111122111111 contains a 1122 block surrounded by two symmetric sub-blocks:

$$22\ \underbrace{1111221111}\ 11\ =\ 22\ \underbrace{\overline{11}\ 1122\ \overline{1111}}\ 11,$$

the same is not the case for 221122221111. Let us explore what happens if a large symmetric block is composed of two smaller symmetric blocks separated by a 1122 block. That is, suppose we have

$$x\ =\ [\overline{2;2},\ a_1, a_2,\ \ldots,\ a_p, a_p,\ \ldots,\ a_2, a_1,\ \overline{1,1}]$$

with $x = (Ax + A')/(Bx + B')$, where $A + B' = 3B$ and $AB' - A'B = 1$ as in the proof of Lemma 7, and

$$y\ =\ [\overline{2;2},\ b_1, b_2,\ \ldots,\ b_q, b_q,\ \ldots,\ b_2, b_1,\ \overline{1,1}]$$

with $y = (\mathcal{A}y + \mathcal{A}')/(\mathcal{B}y + \mathcal{B}')$, where $\mathcal{A} + \mathcal{B}' = 3\mathcal{B}$ and $\mathcal{A}\mathcal{B}' - \mathcal{A}'\mathcal{B} = 1$, such that

$$z\ =\ [\overline{2;2},\ c_1, c_2,\ \ldots,\ c_r, c_r,\ \ldots,\ c_2, c_1,\ \overline{1,1}]$$

is actually

$$z = [\overline{2;2},\ a_1,\ \ldots,\ a_p, a_p,\ \ldots,\ a_1,\ 1, 1, 2, 2,\ b_1,\ \ldots,\ b_q, b_q,\ \ldots,\ b_1,\ \overline{1,1}].$$

Then

$$z = \cfrac{A\left(\dfrac{\mathcal{A}z + \mathcal{A}'}{\mathcal{B}z + \mathcal{B}'}\right) + A'}{B\left(\dfrac{\mathcal{A}z + \mathcal{A}'}{\mathcal{B}z + \mathcal{B}'}\right) + B'} = \frac{(A\mathcal{A} + A'\mathcal{B})z + (A\mathcal{A}' + A'\mathcal{B}')}{(B\mathcal{A} + B'\mathcal{B})z + (B\mathcal{A}' + B'\mathcal{B}')}$$

has $(A\mathcal{A} + A'\mathcal{B}) + (B\mathcal{A}' + B'\mathcal{B}') = 3(B\mathcal{A} + B'\mathcal{B})$, which simplifies to $A'\mathcal{B} + B\mathcal{A}' = 2B'\mathcal{A}$, and $(A\mathcal{A} + A'\mathcal{B})(B\mathcal{A}' + B'\mathcal{B}') - (A\mathcal{A}' + A'\mathcal{B}')(B\mathcal{A} + B'\mathcal{B}) = 1$. How is the corresponding quadratic equation for z related to those for x and y? Since the leading coefficients are $(B\mathcal{A} + B'\mathcal{B})$, B and $\mathcal{B}$, we find that

$$
\begin{aligned}
(B\mathcal{A} + B'\mathcal{B})^2 &= B^2(\mathcal{A})^2 + B\mathcal{B}(2B'\mathcal{A}) + (B')^2\mathcal{B}^2 \\
&= B^2\big(\mathcal{A}(3\mathcal{B} - \mathcal{B}')\big) + B\mathcal{B}(A'\mathcal{B} + B\mathcal{A}') + \big(B'(3B - A)\big)\mathcal{B}^2 \\
&= B^2\big(3\mathcal{A}\mathcal{B} - (\mathcal{A}\mathcal{B}' - \mathcal{A}'\mathcal{B})\big) + \mathcal{B}^2\big(3BB' - (AB' - A'B)\big) \\
&= -B^2 - \mathcal{B}^2 + 3B\mathcal{B}(B\mathcal{A} + B'\mathcal{B})
\end{aligned}
$$

and so

$$(B\mathcal{A} + B'\mathcal{B})^2 + B^2 + \mathcal{B}^2 = 3B\mathcal{B}(B\mathcal{A} + B'\mathcal{B}). \tag{12}$$

A *Markov triple* (m, m_1, m_2) consists of three positive integers such that

$$m^2 + m_1^2 + m_2^2 = 3mm_1m_2 \tag{13}$$

and a *Markov number* is an integer that belongs to (at least one) Markov triple. If two of the numbers in a triple are the same, say $m_1 = m_2$, then (13) becomes $m^2 + 2m_1^2 = 3mm_1^2$, and so m_1 is a divisor of m. Writing $m = nm_1$ and dividing by m_1^2, we obtain $n^2 + 2 = 3nm_1$, and thus n must be 1 or 2 and m_1 must be 1. Hence the only triples with at least two equal terms are $(1,1,1)$ and $(2,1,1)$, with its permutations. These solutions of (13) are called the *singular solutions*. The *non-singular solutions* thus consist of three distinct Markov numbers.

Taking m_1 and m_2 as given, equation (13) becomes a quadratic equation in m,

$$m^2 - (3m_1m_2)m + (m_1^2 + m_2^2) = 0,$$

and the other root $m' = 3m_1m_2 - m$ of this equation is also a positive integer, since $mm' = m_1^2 + m_2^2$. Further, if $m_1 > m_2$,

$$(m_1 - m)(m_1 - m') = m_1^2\Big(2 + \Big(\frac{m_2}{m_1}\Big)^2 - 3m_2\Big) < 0,$$

so that m_1 is strictly between m' and m (similarly if $m_2 > m_1$). Thus every non-singular solution (m, m_1, m_2) has three distinct *neighbors*,

$$(m', m_1, m_2), \ (m, m_1', m_2) \text{ and } (m, m_1, m_2')$$

where

$$m' = 3m_1m_2 - m, \ m_1' = 3mm_2 - m_1 \text{ and } m_2' = 3mm_1 - m_2;$$

up to permutations, the only neighbor of the singular solution $(1,1,1)$ is $(2,1,1)$ and the only neighbors of $(2,1,1)$ are $(1,1,1)$ and $(5,1,2)$. Writing the non-singular solution (m, m_1, m_2) so that $m > \max(m_1, m_2)$, we have that $m' < \max(m_1, m_2) < m$ while $m_1 < m < m_1'$ and $m_2 < m < m_2'$. Thus two of the neighbors have bigger largest members and the other has a smaller largest member. Starting with any given non-singular solution, we may now pass through a sequence of successively smaller neighbors to arrive at the singular solution $(1,1,1)$. This reduction process also establishes that the numbers in any Markov triple are pairwise relatively prime, because their greatest common divisor must also be a divisor of the members of each smaller neighbor and hence a divisor of 1.

Since the members of a non-singular solution (m, m_1, m_2) are pairwise relatively prime, we can find integers k, k_1 and k_2 such that

$$m_1k \equiv m_2 \bmod m, \ m_2k_1 \equiv m \bmod m_1 \text{ and } mk_2 \equiv m_1 \bmod m_2 \qquad (14)$$

with $0 < k < m$, $0 < k_1 < m_1$ (unless $m_1 = 1$, in which case we take $k_1 = 0$) and $0 < k_2 < m_2$ (unless $m_2 = 1$, in which case we take $k_2 = 1$). Since $m_1^2 + m_2^2 \equiv 0 \bmod m$ by (13) and, similarly, for $m^2 + m_2^2$ and $m^2 + m_2^2$, we also have for k, k_1 and k_2 that

$$m_2k \equiv -m_1 \bmod m, \ mk_1 \equiv -m_2 \bmod m_1 \text{ and } m_1k_2 \equiv -m \bmod m_2. \qquad (15)$$

Further, since the bigger neighbors of (m, m_1, m_2) have $m_1' \equiv -m_1 \bmod m$ or m_2 and $m_2' \equiv -m_2 \bmod m$ or m_1, the *complete Markov triple* $(m, k; m_1, k_1; m_2, k_2)$ naturally extends to the bigger neighbors $(m_1', k_1'; m, k; m_2, k_2)$ and $(m_2', k_2'; m, k; m_1, k_1)$. We shall make use of the following relations between the m's and k's:

$$m_1 = mk_2 - km_2, \ m_2 = km_1 - mk_1 \text{ and } m' = m_1 k_2 - k_1 m_2. \tag{16}$$

For the formula for m_1, we have from (14) and (15) that $mk_2 - km_2 \equiv m_1 \bmod$ both m_2 and m, and so $mk_2 - km_2 \equiv m_1 \bmod mm_2$, since m and m_2 are relatively prime. However,

$$(mk_2 - km_2) - m_1 < mk_2 \leq mm_2$$

and

$$(mk_2 - km_2) - m_1 \geq \big(m - (m-1)m_2\big) - m_1 > -mm_2,$$

so that $(mk_2 - km_2) - m_1 = 0$ as desired. The other parts of (16) follow by similar arguments.

We connect the solutions of (13) to forms of the type described in Lemma 8 as follows. To each Markov triple (m, m_1, m_2) we can associate a form

$$mx^2 + (3m - 2k)xy + (l - 3k)y^2$$

where, as before, $m_1 k \equiv m_2 \bmod m$ satisfies $0 < k < m$ (unless $m = 1$ in which case $k = 0$) and $k^2 + 1 = lm$. Since the triple could either be (m, m_1, m_2) or (m, m_2, m_1), we now will order the members of the triple so that we obtain a form with $0 < 2k < m$ as in Lemma 8. For a non-singular solution (m, m_1, m_2), let $m_1 k \equiv m_2 \bmod m$ and $m_2 k^* \equiv m_1 \bmod m$ be $0 < k, k^* < m$. We also have $m_2 k \equiv -m_1 \bmod m$ by (15), so $m_2(k + k^*) \equiv 0 \bmod m$ and $k^* = m - k$. Thus we may choose the triple (m, m_1, m_2) where $m \geq \max(m_1, m_2)$ to also have $0 \leq 2k \leq m$. With this understanding, we define the *Markov form* $f_m(x, y)$ corresponding to (m, m_1, m_2) to be

$$f_m(x, y) \ = \ mx^2 + (3m - 2k)xy + (l - 3k)y^2,$$

where $m_1 k \equiv m_2 \bmod m$ satisfies $0 \leq 2k \leq m$ and $k^2 + 1 = lm$. For a non-singular solution, we note that the form $f_m^*(x,y)$ defined using $k^* = m - k$ is equivalent to $f_m(x,y)$ since $f_m^*(x,y) = f_m(x + 2y, -y)$. Thus $(1,1,1)$ gives

$$f_1(x,y) = x^2 + 3xy + y^2,$$

which is equivalent to $x^2 + xy - y^2 = f_1(x + 2y, -y)$ and corresponds to (1), and $(2,1,1)$ gives

$$f_2(x,y) = 2x^2 + 4xy - 2y^2,$$

which is twice $x^2 + 2xy - y^2$ and corresponds to (2). The non-singular solutions give reduced forms as in Lemma 8 that by (13) should satisfy the symmetry (12) suggested by the examples. Figure 1 shows part of the *Markov chain* of complete Markov triples, where each is connected to its neighbors, and we also have given the sequence of 1's and 2's making up the purely periodic continued fraction that is a root of $f_m(x,1) = 0$.

Lemma 9. *Let x and y be integers. Then* $\min_{(x,y) \neq (0,0)} |f_m(x,y)| = m$.

Proof. We first remark that $f_m(x,y) \sim -f_m(x,y)$. For the singular solutions, we have $f_1(x + 2y, -x - y) = -f_1(x,y)$ and $f_2(y, -x) = -f_2(x,y)$, while for the non-singular solutions, we have the identity

$$mm_1 f_m(k_1 x - l_1 y, m_1 x - k_1 y) = -mm_1 f_m(x,y), \qquad (17)$$

where k_1 is as in (14) and $k_1^2 + 1 = l_1 m_1$, because $m_2 = km_1 - mk_1$ by (16). Since the two roots of $f_m(x,y)$ and of $-f_m(x,y)$ are the same up to order, the equivalence between forms given in (17) shows that the roots of $f_m(x,1) = 0$ are equivalent as real numbers with

$$\theta = \frac{k_1 \varphi - l_1}{m_1 \varphi - k_1}.$$

As it will be easier to work with $mf_m(x,y)$ instead of $f_m(x,y)$ itself, we rewrite

$$mf_m(x,y) = m^2 x^2 + m(3m - 2k)xy + m(l - 3k)y^2$$

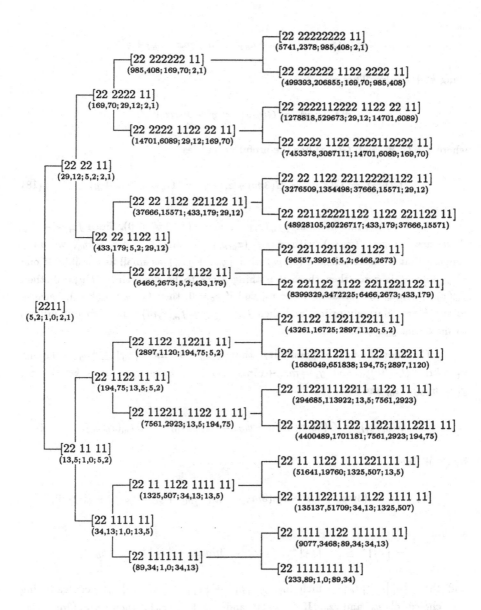

Figure 1. The first 31 non-singular members of the Markov chain.

as

$$mf_m(x,y) = (mx - ky)^2 + 3my(mx - ky) + y^2$$

using $k^2 + 1 = lm$. That is,

$$mf_m(x,y) = G(y,z) = y^2 + 3myz + z^2,$$

where $z = (mx - ky)$. $G(y,z)$ has the useful properties

$$G(y,z) = G(z,y) = G(3my + z, -y) = G(-z, 3mz + y). \qquad (18)$$

Let q be the minimum of $|f_m(x,y)|$, where $(x,y) \neq (0,0)$. Since $f_m \sim -f_m$, there are integers (x_0, y_0) with $q = f_m(x_0, y_0)$. Setting $z_0 = mx_0 - ky_0$, we may suppose that we have selected (x_0, y_0) with $|y_0| + |z_0|$ as small as possible. If one of y_0 and z_0 is zero, then there is nothing more to be done since if $y_0 = 0$, then $f_m(x_0, 0) = mx_0^2$ has smallest value m, and if $z_0 = 0$, then $(x_0, y_0) = (k, m)$, because m and k are relatively prime, and then $f_m(x_0, y_0) = f_m(k, m) = m$. Hence it suffices to show that $y_0 z_0 = 0$.

Consider first the possibility that $y_0 z_0 < 0$ and $|y_0| \leq |z_0|$. Taking $x_1 = kx_0 - (l - 3k)y_0$ and $y_1 = mx_0 + (3m - k)y_0$, we have $z_1 = mx_1 - ky_1 = -y_0$, $y_1 = 3my_0 + z_0$, and thus

$$G(y_1, z_1) = G(3my_0 + z_0, -y_0) = G(y_0, z_0) = mf_m(x_0, y_0),$$

by (18). But

$$G(y_1, z_1) = (3my_0 + z_0)^2 + 3m(3my_0 + z_0)(-y_0) + y_0^2 = y_1 z_0 + y_0^2 > 0,$$

so that

$$-|z_0|^2 \leq -|y_0|^2 < y_1 z_0 = 3my_0 z_0 + |z_0|^2 < |z_0|^2,$$

and thus $|y_1| < |z_0|$, which means $|y_1| + |z_1| < |y_0| + |z_0|$, contradicting the choice of y_0 and z_0. If $y_0 z_0 < 0$ and $|y_0| > |z_0|$, then $x_2 = (3m - k)x_0$

$+ (l - 3k)y_0$ and $y_2 = -mx_0 + ky_0$ make $z_2 = 3mz_0 + y_0$ and $y_2 = -z_0$, and we obtain a similar contradiction. Thus we must have $y_0 z_0 \geq 0$. Let $x_3 = kx_0 - ly_0$ and $y_3 = mx_0 - ky_0$ so that $z_3 = -y_0$ and $y_3 = z_0$. Then

$$G(y_3, z_3) \;=\; G(z_0, -y_0) \;=\; y_0^2 - 3my_0 z_0 + z_0^2$$

gives

$$| G(y_3, z_3) | \;\leq\; | y_0^2 + z_0^2 | + | -3my_0 z_0 | \;=\; G(y_0, z_0)$$

and so, because $G(y_0, z_0)$ is the minimal value of $mf_m(x, y)$, we must have $| y_0^2 + z_0^2 - 3my_0 z_0 | = | y_0^2 + z_0^2 + 3my_0 z_0 |$. So $y_0 z_0 = 0$ and we are finished.

Since $d(f_m) = 9m^2 - 4$, we immediately have that

$$\frac{\mu(f_m)}{\sqrt{d(f_m)}} \;=\; \frac{m}{\sqrt{9m^2 - 4}} \;>\; \frac{1}{3}$$

for every Markov number m. We conclude with a result of Markov that completely determines the Markov and Lagrange spectra above 1/3.

Theorem 4. *The Markov and Lagrange spectra above 1/3 consist of the numbers $m/\sqrt{9m^2 - 4}$, where m is a Markov number.*

As in the proof of Lemma 9, we will simplify our calculations by considering a multiple of $f_m(x, y)$ instead of the Markov form itself. Let $mF_m(x, y) = f_m(x, y)$ so that

$$F_m(x, y) \;=\; x^2 + \beta_m xy + \gamma_m y^2,$$

where $\beta_m = (3m - 2k)/m$ and $\gamma_m = (l - 3k)/m$, and we have that $2 \leq \beta_m \leq 3$, $d(F_m) = 9 - 4/m^2$ and min $| F_m(x, y) | = 1$, where $(x, y) \neq (0, 0)$. Since $\mu(f)/\sqrt{d(f)}$ is unchanged if f is replaced by a non-zero multiple, we will also suppose that $f(x, y) = \alpha x^2 + \beta xy + \gamma y^2$ has been "normalized" to have $\alpha = 1$, so that $f(x, y) = x^2 + \beta xy + \gamma y^2$. The proof of Theorem 4 will proceed by comparing the values of $f(x, y)$ and $F_m(x, y)$ at four points generated from the Markov number m. In the next three lemmas, $f(x, y)$ denotes a form $x^2 + \beta xy + \gamma y^2$ and $F_m(x, y)$ is as above.

Lemma   10. $F_m(k, m) = 1,$ $F_m(k - 3m, m) = 1,$ $F_m(k_1, m_1) = -1$ *and*
$F_m(k_2 - 3m_2, m_2) = -1.$

Proof. The corresponding identities for $m^2 F_m = \pm m^2$ follow from the definitions of the k's in (14) and the relations in (16).

Lemma 11. *If* $f(x, y)$ *satisfies* $f(k, m) \geq 1,$ $f(k - 3m, m) \geq 1,$ $f(k_1, m_1) \leq -1$ *and* $f(k_2 - 3m_2, m_2) \leq -1,$ *then* $f = F_m.$

Proof. The inequalities for $f(x, y)$ and the values of Lemma 10 for $F_m(x, y)$ combine to give four relations between f and F_m:

$$km\beta + m^2\gamma \;\geq\; km\beta_m + m^2\gamma_m \tag{19}$$

$$-(3m - k)m\beta + m^2\gamma \;\geq\; -(3m - k)m\beta_m + m^2\gamma_m \tag{20}$$

$$k_1 m_1\beta + m_1^2\gamma \;\leq\; k_1 m_1\beta_m + m_1^2\gamma_m \tag{21}$$

$$-(3m_2 - k_2)m_2\beta + m_2^2\gamma \;\leq\; -(3m_2 - k_2)m_2\beta_m + m_2^2\gamma_m. \tag{22}$$

Combining $(3m - k)$ times (19) with k times (20), we have $\gamma \geq \gamma_m$, while $(3m_2 - k_2)m_2$ times (21) together with $k_1 m_1$ times (22) gives $\gamma \leq \gamma_m$. Thus $\gamma = \gamma_m$, (19) becomes $\beta \geq \beta_m$, (20) becomes $\beta \leq \beta_m$, and thus $\beta = \beta_m$. Hence $f(x, y) = F_m(x, y)$.

Lemma 12. *If* $f(k_1, m_1) \leq -1$ *and* $f(k_2 - 3m_2, m_2) \leq -1,$ *then* $d(f) \geq d(F_m).$

Proof. By completing the squares, we have $f = (x + \beta y/2)^2 - \Delta y^2$ and $F_m = (x + \beta_m y/2)^2 - \Delta_m y^2$, where $\Delta = d(f)/4$ and $\Delta_m = d(F_m)/4$. It suffices to show that $0 \leq \Delta - \Delta_m$. By Lemma 10, $f(k_1, m_1) \leq F_m(k_1, m_1)$ and so

$$\left(\frac{k_1}{m_1} + \frac{\beta}{2}\right)^2 - \left(\frac{k_1}{m_1} + \frac{\beta_m}{2}\right)^2 \;\leq\; \Delta - \Delta_m.$$

Since $k_1/m_1 > 0$ and $\beta_m > 0$, we have $0 \leq \Delta - \Delta_m$ provided $\beta \geq \beta_m$. If $\beta < \beta_m$, then $f(k_2 - 3m_2, m_2) \leq F_m(k_2 - 3m_2, m_2)$ yields

$$\left(\frac{3m_2 - k_2}{m_1} - \frac{\beta}{2}\right)^2 - \left(\frac{3m_2 - k_2}{m_1} - \frac{\beta_m}{2}\right)^2 \leq \Delta - \Delta_m$$

and $0 \leq \Delta - \Delta_m$ again follows, since $3m_2 - k_2 \geq 2m_2$ yet $\beta_m/2 < 2$.

Lemma 13. *If $f(k,m) \leq -1$ and $f(k-3m,m) \leq -1$, then $d(f) > 9$.*

Proof. Proceeding as in the previous proof, we have from $f(k,m) \leq F_m(k,m) - 2$ that

$$\left(\frac{k}{m} + \frac{\beta}{2}\right)^2 - \left(\frac{k}{m} + \frac{\beta_m}{2}\right)^2 \leq \Delta - \Delta_m - \frac{2}{m^2}$$

and then $\Delta - \Delta_m \geq 2/m^2$ if $\beta \geq \beta_m$ and hence $d(f) \geq d(F_m) + 8/m^2 > 9$. If $\beta < \beta_m$, then $f(k-3m,m) \leq F_m(k-3m,m) - 2$ again yields $\Delta - \Delta_m \geq 2/m^2$, and we are finished.

Proof of Theorem 4. Since we are interested only in $\mu(f)/\sqrt{d(f)}$, we may suppose by Lemma 3 that $|\beta| \leq |\alpha|$, and then, by dividing through by $|\alpha|$, we may suppose that $\alpha = 1$ and $|\beta| \leq 1$. If $\beta \geq 0$, then $f(x+y,y)$ gives an equivalent form with $\alpha = 1$ and $2 \leq \beta \leq 3$, while if $\beta < 0$, $f(x+y, -y)$ satisfies $\alpha = 1$ and $2 \leq \beta \leq 3$. Thus it suffices to show that any form $f(x,y) = x^2 + \beta xy + \gamma y^2$ such that $2 \leq \beta \leq 3$, $0 < d(f) < 9$ and $|f(x,y)| \geq 1$ for integers $(x,y) \neq (0,0)$, is in fact an $F_m(x,y)$ for some Markov number m. By Lemma 13, we can never have $f(k,m) \leq -1$ and $f(k-3m,m) \leq -1$, while Lemma 12 may be regarded as explaining that if Lemma 11 has failed to make $f = F_m$, then we should try a larger m.

Let $f(x,y)$ be a form as above. If $f(1,-1) \geq 1$, then $-\beta + \gamma \geq 0$, which contradicts the conditions $2 \leq \beta \leq 3$ and $0 < d(f) < 9$. Thus we must have $f(1,-1) \leq -1$ and so $-\beta + \gamma \leq -2$. If $f(0,1) \geq 1$, then $\gamma \geq 1$ forces $\beta \geq 3$. Thus $\beta = 3$ and $\gamma = 1$, and we have $f(x,y) = x^2 + 3xy + y^2$, which is the first Markov form. Otherwise, $f(0,1) \leq -1$ and so $\gamma \leq -1$. If $f(-5,2) \geq 1$, then $25 - 10\beta + 4\gamma \geq 1$, and thus $10\beta \leq 24 + 4\gamma \leq 20$, since $\gamma \leq -1$. But $2 \leq \beta \leq 3$ forces $\beta = 2$ and $\gamma = -1$, and we have $f(x,y) = x^2 + 2xy - y^2$, which is half the second Markov form. Otherwise, we have $f(0,1) \leq -1$ and $f(-5,2) \leq -1$. Regarding $(0,1)$ and $(-5,2)$ as (k_1,m_1) and $(k_2 - 3m_2, m_2)$ from the third complete Markov triple $(m,k;$

$m_1, k_1; \ m_2, k_2) = (5, 2; \ 1, 0; \ 2, 1)$, we can now consider the values of $f(k, m)$ and $f(k - 3m, m)$. If these are both positive, then f is F_m by Lemma 11. If not, they can not both be negative by Lemma 13 and we must have either

$$f(k, m) \leq -1 \quad \text{and} \quad f(k_2 - 3m_2, m_2) \leq -1$$

or

$$f(k_1, m_1) \leq -1 \quad \text{and} \quad f(k - 3m, m) \leq -1.$$

But these possibilities are just

$$f(k_1, m_1) \leq -1 \quad \text{and} \quad f(k_2 - 3m_2, m_2) \leq -1$$

for the two bigger neighbors $(m_1', k_1'; \ m, k; \ m_2, k_2)$ and $(m_2', k_2'; \ m, k; \ m_1, k_1)$ of our current triple $(m, k; \ m_1, k_1; \ m_2, k_2)$. If this process does not ultimately have $f = F_m$ for some Markov number m, then, by Lemma 12, $d(f)$ would be greater than $9 - 4/m^2$ for every m, which contradicts the assumption that $d(f) < 9$.

§7. Asymmetric approximation and Segre's theorem.

From Hurwitz's theorem, we know that the symmetric approximation $b \, | \, bt - a \, | \, < K$ will have infinitely many solutions for any t only if $K \geq 1/\sqrt{5}$. But we also know that the convergents of t approximate t alternately from below and from above and that while the approximation may be bad on one side, it may be quite good on the other. Segre [1945] introduced the idea of asymmetric approximations in his study of lattice points via geometry of numbers. We shall use the method of Theorem 5.2 to show his principal result.

Theorem 1. *Let t be any irrational number and let $0 < K \leq 1$ be a given constant. Then at least one of the inequalities*

$$\begin{cases} \dfrac{-1}{\sqrt{1 + 4K}} \ \leq \ b(bt - a) \ < \ 0 \\[2ex] 0 \ < \ b(bt - a) \ \leq \ \dfrac{K}{\sqrt{1 + 4K}} \end{cases}$$

has an infinity of solutions, where $b > 0$ and a are integers.

Proof. Since we have from (I.2.6) that

$$B_k(B_k t - A_k) = \frac{(-1)^k}{\zeta_{k+1} + \xi_k},$$

it suffices to show that for an infinity of odd k's at least one of the following holds:

$$\begin{aligned}
\zeta_{k+1} + \xi_k &\leq \sqrt{1+4K} \\
\zeta_{k+2} + \xi_{k+1} &\leq \frac{\sqrt{1+4K}}{K} \\
\zeta_{k+3} + \xi_{k+2} &\leq \sqrt{1+4K}.
\end{aligned} \tag{1}$$

Suppose first that for all sufficiently large k's the partial quotients a_{k+2} are bounded by $1/K$, so that $a_{k+2} < 1/K$. Using the continued fractions (I.1.1) with real valued partial quotients, we may write

$$\zeta_{k+1} > [1; \overline{1/K, 1}] \quad \text{and} \quad \xi_k > [0; \overline{1/K, 1}].$$

Let $x = [0; \overline{1/K, 1}]$; that is,

$$x = \cfrac{1}{\cfrac{1}{K} + \cfrac{1}{1+x}}$$

and so

$$x^2 + x - K = 0 \quad \text{and} \quad x > 0.$$

Then

$$\zeta_{k+1} + \xi_k > (1+x) + x = 1 + (-1 + \sqrt{1+4K}) = \sqrt{1+4K}$$

and we are done.

Now suppose that for an infinity of odd k's we have $a_{k+2} \geq 1/K$. As in the proof of Theorem 5.2, we look at the first two inequalities of (1) and then at the second and third. From the first two, we obtain

$$1 < \left(\sqrt{1+4K} - \frac{1}{\xi_{k+1}}\right)\left(\frac{\sqrt{1+4K}}{K} - \xi_{k+1}\right)$$

and so

$$K \xi_{k+1}^2 - \sqrt{1+4K}\, \xi_{k+1} + 1 < 0, \tag{2}$$

while from the second and third we find that

$$\xi_{k+2}^2 - \sqrt{1+4K}\, \xi_{k+2} + K \; < \; 0. \tag{3}$$

From (2), we have that $\xi_{k+1} > (\sqrt{1+4K}-1)/(2K)$ while from (3), $\xi_{k+2} > (\sqrt{1+4K}-1)/2$. But

$$\xi_{k+2} \;=\; \frac{1}{a_{k+2}+\xi_{k+1}} \;<\; \frac{1}{\dfrac{1}{K}+\dfrac{\sqrt{1+4K}-1}{2K}} \;=\; \frac{\sqrt{1+4K}-1}{2},$$

and this contradiction completes the proof.

§8. Approximation by non-convergents.

We have seen that the convergents of the irrational number t give the solutions of $b\,\|\,bt\,\| < K$ when K is a small positive constant and, in §6, we explored the relationship between t and K in detail for $K \le 1/\sqrt{5}$. In §5 of Chapter II, we saw that the solutions were multiples of convergents for $K \le 1/2$, either multiples of convergents or "short" Ostrowski sums of the form $B_k + c_{k+2}B_{k+1}$ for $K = 1$, and then (in Theorem II.5.3) longer Ostrowski sums became possible for larger values of K. However, the Corollary to Theorem II.5.2 shows that the number e is such that when $K = 1$ the only "good" approximations are multiples of convergents. In this section we consider how "bad" the approximation by non-convergents can be.

The "shortest" non-convergents are the "intermediate" convergents of the form $b = B_k + c_{k+2}B_{k+1}$ where $1 \le c_{k+2} < a_{k+2}$. For such numbers,

$$b\,\|\,bt\,\| \;=\; \Big(B_k + c_{k+2}B_{k+1}\Big)\Big|D_k + c_{k+2}D_{k+1}\Big|$$

$$=\; \Big(B_k + c_{k+2}B_{k+1}\Big)\big|D_k\big|\Big(1 - \frac{c_{k+2}}{\zeta_{k+2}}\Big) \tag{1}$$

and the minimum of this quadratic expression in c_{k+2} must occur either when $c_{k+2} = 1$ or when $c_{k+2} = a_{k+2} - 1$.

Lemma 1. *If* $a_{k+2} > 1$, *then* $(B_k + B_{k+1}) \| (B_k + B_{k+1})t \| < 2$.

Proof. Since $c_{k+2} = 1$ and $B_k + B_{k+1} = (a_{k+1} + 1)B_k + B_{k-1}$, we have by (1) that

$$(B_k + B_{k+1}) \| (B_k + B_{k+1})t \| = \frac{\big((a_{k+1} + 1) + \xi_k\big)\big(\zeta_{k+2} - 1\big)}{(a_{k+1} + \xi_k)\zeta_{k+2} + 1}.$$

But $a_{k+1} + \xi_k + 1 < 2(a_{k+1} + \xi_k)$ and the result follows.

Thus if t has infinitely many intermediate convergents, then there is an infinite number of non-convergent solutions of $b \| bt \| < 2$. If t has only a finite number of intermediate convergents, then t is equivalent to $g = [1; 1, 1, \dots]$. For such a number and with k sufficiently large, $B_k + B_{k+2}$ and $A_k + A_{k+2}$ are relatively prime, since

$$(B_k + B_{k+2})A_{k+2} - (A_k + A_{k+2})B_{k+2}$$
$$= (2B_k + B_{k+1})(A_k + A_{k+1}) - (2A_k + A_{k+1})(B_k + B_{k+1}) = (-1)^k,$$

while $B_k + B_{k+3} = 2B_{k+2}$ is a multiple of a convergent.

Theorem 1. *Let* t *be equivalent to* g. *Then for any* $\epsilon > 0$ *there are at most a finite number of relatively prime non-convergent solutions of* $b \| bt \| < \sqrt{5} - \epsilon$.

Proof. Let $b = B_k + B_{k+2}$ for k sufficiently large. Then

$$b \| bt \| = (B_k + B_{k+2}) |D_k| \left(1 + \frac{1}{\zeta_{k+2}\zeta_{k+3}} \right)$$
$$= \frac{3 + \xi_k}{\zeta_{k+1} + \xi_k} \left(1 + \frac{1}{\zeta_{k+2}\zeta_{k+3}} \right).$$

Since $\xi_k = [0; 1, 1, \dots, 1, \dots, a_1] \to 1/g$ as $k \to \infty$, we see that

$$b \| bt \| \to \frac{3 + \frac{1}{g}}{g + \frac{1}{g}} \left(1 + \frac{1}{g^2} \right) = \sqrt{5}$$

as $k \to \infty$ and we are done.

Combining this result with the lemma, we obtain:

Corollary. *Let t be an irrational number and let b be a positive integer not of the form cB_k for some index k and positive integer c. Then there are infinitely many solutions b of $b \parallel bt \parallel < \sqrt{5}$ and the constant $\sqrt{5}$ is the best possible.*

To investigate those numbers that are not equivalent to g, we first show:

Lemma 2. *If $a_{k+1} = 1$ and $a_{k+2} > 2$, then*

$$(B_k + B_{k+1}) \parallel (B_k + B_{k+1})t \parallel$$

$$< \left(B_k + (a_{k+2} - 1)B_{k+1}\right) \parallel \left(B_k + (a_{k+2} - 1)B_{k+1}\right)t \parallel \qquad (2)$$

if and only if $1 + \xi_k > \zeta_{k+3}$.

Proof. Since $a_{k+1} = 1$, $B_{k+1} = B_k + B_{k-1}$ and we may apply (1) to find

$$(B_k + B_{k+1}) \parallel (B_k + B_{k+1})t \parallel \; = \; \frac{(2 + \xi_k)(\zeta_{k+2} - 1)}{\zeta_{k+2}(1 + \xi_k) + 1}$$

and

$$\left(B_k + (a_{k+2} - 1)B_{k+1}\right) \parallel \left(B_k + (a_{k+2} - 1)B_{k+1}\right)t \parallel$$

$$= \frac{a_{k+2}(1 + \xi_k) - \xi_k}{\zeta_{k+2}(1 + \xi_k) + 1} \left((\zeta_{k+2} - a_{k+2}) + 1 \right).$$

Thus (2) is equivalent to

$$(2 + \xi_k)(\zeta_{k+2} - 1) \; < \; \left(a_{k+2}(1 + \xi_k) - \xi_k\right)\left((\zeta_{k+2} - a_{k+2}) + 1\right)$$

and, since $\zeta_{k+2} = a_{k+2} + 1/\zeta_{k+3}$, this is the same as

$$2\left((1 + \xi_k)(1 + \frac{1}{\zeta_{k+3}}) - (2 + \xi_k)\right) \; < \; a_{k+2}\left((1 + \xi_k)(1 + \frac{1}{\zeta_{k+3}}) - (2 + \xi_k)\right).$$

But $a_{k+2} > 2$, so this is equivalent to

$$\left(1+\xi_k\right)\!\left(1+\frac{1}{\zeta_{k+3}}\right) \;>\; 2+\xi_k,$$

which is the same as what was to be shown.

From Lemma 1, the intermediate convergents will give infinitely many relatively prime non-convergent solutions to $b \,\|\, bt \,\| \,< 2$ for any t not equivalent to g. We now use Lemma 2 to exhibit numbers for which 2 is the best possible constant.

Theorem 2. *Let t be equivalent to a number such that $a_k \to \infty$ for $k \equiv 0 \bmod 2$ and $a_k = 1$ for $k \equiv 1 \bmod 2$. Then for any $\epsilon > 0$ there are at most a finite number of relatively prime non-convergent solutions of $b \,\|\, bt \,\| \,< 2 - \epsilon$.*

Proof. Let $k \equiv 0 \bmod 2$ be sufficiently large. Then $1 + \xi_k = [1; a_k, 1, \dots, a_1]$ and $\zeta_{k+3} = [1; a_{k+4}, 1, \dots]$. By Lemma 2 it suffices to consider only $b = B_k + B_{k+1}$, in which case we have

$$b \,\|\, bt \,\| \;=\; \frac{2+\xi_k}{1+\xi_k+\dfrac{1}{\zeta_{k+2}}}\left(1 - \frac{1}{\zeta_{k+2}}\right).$$

But $\zeta_{k+2} \to \infty$ and $\xi_k = [0; a_k + \xi_{k-1}] \to 0$ as $k \to \infty$, and so $b \,\|\, bt \,\| \to 2$.

It now is clear that $b \,\|\, bt \,\| \,< K$ will have infinitely many positive integer solutions for $K < 2$ provided there are infinitely many indices k either with $a_k > 1$ and a_{k+1} arbitrarily large or with $a_k = 1$, $a_{k+1} = 1$ and a_{k+2} arbitrarily large. We know from §8 of Chapter I that $e = [2; 1, 2, 1, 1, 4, 1, 1, 6, 1, 1, 8, \dots]$; that is, $a_{3k} = 1$, $a_{3k+1} = 1$ and $a_{3k+2} = 2(k+1)$ for $k \geq 1$. We conclude this section by showing that $K = 3/2$ is the best possible value for "e-like" numbers.

Theorem 3. *Let t be equivalent to a number such that $a_k = 1$ for $k \equiv 0, 1 \bmod 3$ and $a_k \to \infty$ for $k \equiv 2 \bmod 3$. Then for any $\epsilon > 0$ there are at most a finite number of relatively prime non-convergent solutions of $b \,\|\, bt \,\| \,< 3/2 - \epsilon$.*

Proof. Let $k \equiv 0 \bmod 3$ be sufficiently large. Then $1 + \xi_k = [1; 1, a_{k-1}, 1, 1, \ldots, a_1]$ and $\zeta_{k+3} = [1; 1, a_{k+4}, 1, 1, \ldots]$. By Lemma 2 it suffices to consider only

$$b = B_k + (a_{k+2} - 1) B_{k+1},$$

in which case we have

$$
b \| bt \| = \frac{a_{k+2}(1 + \xi_k) - \xi_k}{\zeta_{k+2}(1 + \xi_k) + 1} \left((\zeta_{k+2} - a_{k+2}) + 1 \right)
$$

$$
= \frac{1 - \dfrac{1}{a_{k+2}} \dfrac{\xi_k}{1 + \xi_k}}{1 + \dfrac{1}{a_{k+2}} \left(\dfrac{1}{\zeta_{k+3}} + \dfrac{1}{1 + \xi_k} \right)} \left(1 + \dfrac{1}{\zeta_{k+3}} \right).
$$

But $\zeta_{k+3} = [1; 1, a_{k+5}, \zeta_{k+6}] \to 2$, $\xi_k = [0; 1, a_{k-1} + \xi_{k-2}] \to 1$ and $1/a_{k+2} \to 0$ as $k \to \infty$, and so $b \| bt \| \to 3/2$.

§9. Inhomogeneous approximation.

In our discussion of homogeneous approximation in the previous sections, we made a detailed study of the possible values of $b \| bt \|$, where b denotes a positive integer and t is a given irrational number, and the Ostrowski representation of b played an central role. In §6 of Chapter II we introduced the t-expansion of an arbitrary real number s and remarked that it would be similarly useful in the investigation of the inhomogeneous approximation $|y| \| yt + s \|$. In this section we shall present such an analysis using a uniform method of attack founded on the t-expansion of s. Throughout this section, t will denote a given irrational number and the notations a_k, ζ_k, A_k, B_k and D_k all refer to the continued fraction of t. Further, s will denote a real number with t-expansion as given by Theorem II.6.1. We shall also let Y_k denote the integer

$$Y_k = c_1 B_0 + \cdots + c_{k+1} B_k$$

and let

$$R_k = c_{k+1} D_k + c_{k+2} D_{k+1} + \cdots$$

so that

$$\| -Y_k t + s \| \ = \ |R_{k+1}|$$

and

$$\| (B_{k+1} - Y_k)t + s \| \ = \ |D_{k+1} + R_{k+1}|.$$

We begin with a famous result of Minkowski.

Theorem 1. *If s is not of the form $mt+n$ for integers m and n, then*

$$\liminf_{|b| \to \infty} \ |b| \ \| bt + s \| \ \le \ \tfrac{1}{4},$$

where b ranges over the integers.

Proof. Suppose first that there are infinitely many indices k such that $c_{k+1} = 0$ and $c_{k+2} > 0$. Then

$$| -Y_k | \ \| -Y_k t + s \| \ = \ Y_k |R_{k+1}|$$

$$| B_{k+1} - Y_k | \ \| (B_{k+1} - Y_k)t + s \| \ = \ (B_{k+1} - Y_k) |D_{k+1} + R_{k+1}|$$

and so

$$\left(| -Y_k | \ \| -Y_k t + s \| \right)\left(| B_{k+1} - Y_k | \ \| (B_{k+1} - Y_k)t + s \| \right)$$
$$= \ B_{k+1}^2 \left(\frac{Y_k}{B_{k+1}} \right)\left(1 - \frac{Y_k}{B_{k+1}} \right)|D_k|^2 \left| \frac{R_{k+1}}{D_k} \right|\left(1 - \left| \frac{R_{k+1}}{D_k} \right| \right),$$

since R_{k+1} and D_k have opposite signs. But $x(1-x) \le 1/4$ for $0 \le x \le 1$, so that

$$\left(| -Y_k | \ \| -Y_k t + s \| \right)\left(| B_{k+1} - Y_k | \ \| (B_{k+1} - Y_k)t + s \| \right) \le \left(\tfrac{1}{4} B_{k+1} |D_k| \right)^2$$

and at least one term on the left must be $\le B_{k+1} |D_k|/4$. By (I.4.3), (I.6.2) and (I.2.4),

$$B_{k+1} |D_k| \ = \ \frac{B_{k+1}}{B_k \left(a_{k+1} + \dfrac{1}{\zeta_{k+2}} \right) + B_{k-1}} \ = \ \frac{1}{1 + \dfrac{1}{\zeta_{k+2}} \xi_{k+1}} \ < \ 1$$

and we are done for this case. Otherwise, we must have that $c_{k+1} > 0$ and $c_{k+2} > 0$ for all sufficiently large k's. For any such index k, we have

$$| -Y_k | \, \| -Y_k t + s \| \; = \; Y_k | R_{k+1} |,$$

$$| B_k + B_{k+1} - Y_k | \, \| (B_k + B_{k+1} - Y_k)t + s \|$$

$$= (B_k + B_{k+1} - Y_k) | D_k + D_{k+1} + R_{k+1} |$$

and so, as in the previous case,

$$\left(| -Y_k | \, \| -Y_k t + s \| \right) \left(| B_k + B_{k+1} - Y_k | \, \| (B_k + B_{k+1} - Y_k)t + s \| \right)$$

$$= (B_k + B_{k+1})^2 \left(\frac{Y_k}{B_k + B_{k+1}} \right) \left(1 - \frac{Y_k}{B_k + B_{k+1}} \right)$$

$$\times \; | D_k + D_{k+1} |^2 \left| \frac{R_{k+1}}{D_k + D_{k+1}} \right| \left(1 - \left| \frac{R_{k+1}}{D_k + D_{k+1}} \right| \right)$$

and at least one of these choices for b makes $| b | \, \| bt + s \|$

$$\leq \tfrac{1}{4} (B_k + B_{k+1}) | D_k + D_{k+1} | \; = \; \tfrac{1}{4} B_{k+1} | D_k | \left(1 + \xi_{k+1} \right) \left(1 - \frac{1}{\zeta_{k+2}} \right).$$

Since $B_{k+1} | D_k | < 1$, we need to consider only $(1 + \xi_{k+1})(1 - 1/\zeta_{k+2})$. If $a_k \to \infty$ as $k \to \infty$, then

$$\lim_{k \to \infty} \left(1 + \xi_{k+1} \right) \left(1 - \frac{1}{\zeta_{k+2}} \right) \; = \; 1$$

and we are done, while if not, $\zeta_{k+2} \leq 1 + 1/\xi_{k+1}$ infinitely many times and this is equivalent to having $(1 + \xi_{k+1})(1 - 1/\zeta_{k+2}) \leq 1$.

The constant $1/4$ is the best possible since there are pairs of numbers t and s such that $\liminf | b | \, \| bt + s \| = 1/4$ where $| b | \to \infty$.

Theorem 2. *Let* $t = [a_0; a_1, a_2, \ldots]$ *have increasing even partial quotients and let*

$$s = \sum_{k=0}^{\infty} \left(\frac{a_{k+1}}{2} \right) D_k.$$

Then

$$\liminf_{|b| \to \infty} |b| \, \|bt + s\| = \frac{1}{4}.$$

Proof. Since

$$|R_{k+1}| = \left| \frac{1}{2} \left(a_{k+2} D_{k+1} + a_{k+4} D_{k+3} + \cdots \right) + \frac{1}{2} \left(a_{k+3} D_{k+2} + a_{k+5} D_{k+4} + \cdots \right) \right|$$

$$= \frac{1}{2} |D_k + D_{k+1}|,$$

we see that $|b| \, \|bt + s\|$ is minimized when b is of the form $-Y_k$ or $B_k + B_{k+1} - Y_k$. In the first case, we have (using I.4.4),

$$|-Y_k| \, \|-Y_k t + s\|$$

$$> \left(\frac{1}{2} B_{k+1} \right) \left(\frac{1}{2} |D_k + D_{k+1}| \right) = \frac{1}{4} B_{k+1} |D_k| \left(1 - \frac{1}{\zeta_{k+2}} \right)$$

which $\to 1/4$ as $k \to \infty$, while in the second case,

$$|B_k + B_{k+1} - Y_k| \, \|(B_k + B_{k+1} - Y_k)t + s\|$$

$$> \frac{1}{4} B_{k+1} |D_k| \left(1 + \xi_{k+1} \right) \left(1 - \frac{1}{\zeta_{k+2}} \right)$$

which also $\to 1/4$ as $k \to \infty$.

We may also show that for any irrational number t there is an s that makes the approximation $|b| \, \|bt + s\|$ large.

Theorem 3. *There exists an $s = s(t)$ such that* $\displaystyle \liminf_{|b| \to \infty} |b| \, \|bt + s\| > \frac{1}{32}$.

Proof. If $a_k \leq 4$ for all sufficiently large k's, then let $s = 0$ so that $|b| \, \| bt + s \|$
$= |b| \, \| bt \|$. Since the homogeneous approximation is minimized when $b = B_k$, we
have

$$| B_k | \, \| B_k t \| \;=\; B_k | D_k | \;=\; \frac{1}{\zeta_{k+1} + \xi_k} \;>\; \frac{1}{(4+1)+1} \;=\; \frac{1}{6}.$$

Otherwise, we have $a_k > 4$ infinitely many times and we choose

$$s \;=\; \sum_{k=0}^{\infty} \left[\frac{a_{k+1}}{2} \right] D_k.$$

Suppose first that $a_k \geq 2$ for all sufficiently large k's. Then $|-Y_k| \, \| -Y_k t + s \|$
is

$$\left(\left[\frac{a_{k+1}}{2} \right] B_k + \left[\frac{a_k}{2} \right] B_{k-1} + \cdots \right) \left| \left[\frac{a_{k+2}}{2} \right] D_{k+1} + \left[\frac{a_{k+3}}{2} \right] D_{k+2} + \cdots \right|$$

$$> \left(\left[\frac{a_{k+1}}{2} \right] B_k \right) \left| \left[\frac{a_{k+2}}{2} \right] D_{k+1} + \frac{1}{2} \left(a_{k+3} D_{k+2} + a_{k+5} D_{k+4} + \cdots \right) \right|$$

which is

$$= \left[\frac{a_{k+1}}{2} \right] B_k | D_{k+1} | \left(\left[\frac{a_{k+2}}{2} \right] - \frac{1}{2} \right) = \frac{\left[\frac{a_{k+1}}{2} \right]}{a_{k+1} + \frac{1}{\zeta_{k+2}} + \xi_k} \; \frac{\left[\frac{a_{k+2}}{2} \right] - \frac{1}{2}}{a_{k+2} + \frac{1}{\zeta_{k+3}}}.$$

For large values of both a_{k+1} and a_{k+2} this estimate is near $1/4$, so we must
examine it for small values only. If $a_{k+1} = 3$ and $a_{k+2} = 3$, we have

$$| -Y_k | \, \| -Y_k t + s \| \;>\; \frac{1}{3 + \frac{1}{3} + \frac{1}{2}} \; \frac{1/2}{3 + \frac{1}{2}} \;>\; \frac{1}{27},$$

while similar calculations for other choices for a_{k+1} and $a_{k+2} \geq 2$ give at least
$1/21$. For $b = B_{k+1} - Y_k$, we note that

$$B_{k+1} - Y_k \;\geq\; B_{k+1} - \frac{1}{2} \left(a_{k+1} B_k + a_k B_{k-1} + \cdots \right)$$

$$> B_{k+1} - \frac{1}{2} \left(B_{k+1} + B_k \right) = \frac{1}{2} B_{k+1} \left(1 - \frac{B_k}{B_{k+1}} \right) > \frac{1}{4} B_{k+1},$$

and so $| B_{k+1} - Y_k | \, \| (B_{k+1} - Y_k) t + s \|$ is

$$> \frac{1}{4} B_{k+1} |D_{k+1}| \left(\left[\frac{a_{k+2}}{2} \right] + \frac{1}{2} \right) \geq \frac{1}{8} B_{k+1} |a_{k+2} D_{k+1}|,$$

since $[x/2] + 1/2 \geq x/2$ for $x \geq 2$. Using (I.4.2, .3 and .4), we find that this last expression is

$$= \frac{1}{8} \frac{1}{1 + \frac{1}{\zeta_{k+2}} \xi_{k+1}} \left(1 - \frac{1}{\zeta_{k+2} \zeta_{k+3}} \right) > \frac{1}{8} \frac{1}{1 + \frac{1}{4}} \left(1 - \frac{1}{4} \right) = \frac{3}{40},$$

and we are done with this case, since it suffices to consider $|b| \, \|bt+s\|$ only for $b = -Y_k$ and for $b = B_{k+1} - Y_k$.

We must now consider the possibility that $a_k = 1$ infinitely many times. Since long blocks of zeros for the c_{k+1}'s are now possible in the t-expansion of our choice for s, we must modify s to ruin the close approximation now given by $-Y_k$. Let s' be obtained from s by replacing each block "000" of three consecutive zero c_{k+1}'s by "001" so that s' contains at most two consecutive zero c_{k+1}'s. To estimate $|b| \, \|bt+s'\|$, it suffices to consider what happens at the end of a sequence of added "001" blocks when we encounter the next partial quotient greater than one (since this means a sudden decrease in the absolute size of the current D_k's). At worst, we have for s' the situation

$$s': \qquad \cdots \underbrace{0\ 0\ \overset{\overset{\displaystyle c_{k+1}}{\downarrow}}{1}\ 0\ 0}_{(a_i\text{'s} = 1)}\ \underset{(a_{k+4} > 1)}{\left[\frac{a_{k+4}}{2} \right]} \cdots ,$$

where our diagram indicates only the c_{k+1}'s of the t-expansion of s' and the corresponding a_{k+1}'s of t. Since the denominators $\{B_k\}_{k \geq 0}$ increase slowest for g, we have that

$$Y_k > B_k + B_{k-3} + B_{k-6} + \cdots > B_k \left(1 + \frac{1}{g^3} + \frac{1}{g^6} + \cdots \right) \sim \left(\frac{3 + \sqrt{5}}{4} \right) B_k,$$

and then for k sufficiently large

$$|-Y_k| \, \| -Y_k t + s \| > \left(\frac{3 + \sqrt{5}}{4} \right) B_k |D_{k+3}| \left(\left[\frac{a_{k+4}}{2} \right] - \frac{1}{2} \right).$$

But this is really

$$\left(\frac{3+\sqrt{5}}{4}\right)B_k \,|\, D_k \,| \; \frac{1}{\zeta_{k+2}\zeta_{k+3}} \; \frac{\left(\left[\frac{a_{k+4}}{2}\right]-\frac{1}{2}\right)}{\zeta_{k+4}}$$

$$= \left(\frac{3+\sqrt{5}}{4}\right) \frac{1}{\zeta_{k+1}+\xi_k} \; \frac{1}{\left(1+\frac{1}{\zeta_{k+3}}\right)\zeta_{k+3}} \; \frac{\left(\left[\frac{a_{k+4}}{2}\right]-\frac{1}{2}\right)}{a_{k+4}+\frac{1}{\zeta_{k+5}}}$$

which is

$$> \left(\frac{3+\sqrt{5}}{4}\right) \frac{1}{[1;\,1,\,1,\,a_{k+4}]+[0;\,1,\,1,\,1]} \; \frac{1}{1+[1;\,a_{k+4}]} \; \frac{\left(\left[\frac{a_{k+4}}{2}\right]-\frac{1}{2}\right)}{a_{k+4}+1}.$$

For $a_{k+4}=3$, we have

$$\frac{3+\sqrt{5}}{4} \; \frac{1}{11/7+2/3} \; \frac{1}{7/3} \; \frac{1/2}{4} \; > \; \frac{1}{32}$$

while $a_{k+4}=2$ and $a_{k+4}\geq 4$ give larger values.

We have seen that the freedom to consider both positive and negative values for b allows us to "flip" from a perhaps bad negative choice $-Y_k$ to a perhaps better positive choice $B_{k+1}-Y_k$ or even $B_k+B_{k+1}-Y_k$. We conclude this section with a result of Khintchine on the size of $|b|\,\|bt+s\|$ when b is restricted to positive integers only.

Theorem 4. *Let s be a real number. Then*

$$\liminf_{b\to\infty} b\,\|bt+s\| \;\leq\; \frac{1}{\sqrt{5}},$$

where b now denotes a positive integer.

Thus the homogeneous case $t=g$ and $s=0$ described in Theorem 5.3 is the "worst case" for the inhomogeneous approximation as well.

Proof. We shall estimate $b\,\|bt+s\|$ for $b=B_{k+1}-Y_k$ if c_{k+2} is small and for

$b = B_{k+2} - Y_{k+1}$ and $B_{k+1} + B_{k+2} - Y_{k+1}$ if c_{k+2} is large. If t has large partial quotients, then at least one of these choices for b will give a very small value for $b \parallel bt + s \parallel$, so we begin with t's with large a_k's and work our way down to $t \sim g$.

Case 1: $a_{k+2} \geq 9$ infinitely many times. For such an a_{k+2}, if $c_{k+2} \leq a_{k+2}/3 - 1$, we choose $b = B_{k+1} - Y_k$ and find that

$$\parallel (B_{k+1} - Y_k)t + s \parallel = |D_{k+1} + R_{k+1}|$$

$$< |(c_{k+2}+1)D_{k+1} - D_{k+2}| \leq |\frac{a_{k+2}}{3}D_{k+1} - D_{k+2}|. \tag{1}$$

Using (I.4.4 and I.2.4), we obtain

$$|\frac{a_{k+2}}{3}D_{k+1} - D_{k+2}| = |D_k|\left(\frac{a_{k+2}}{3\zeta_{k+2}} + \frac{1}{\zeta_{k+2}\zeta_{k+3}}\right)$$

$$= |D_k|\left(\frac{a_{k+2}\zeta_{k+3} + 3}{3(a_{k+2}\zeta_{k+3} + 1)}\right), \tag{2}$$

which is a positive, monotone decreasing function of $\zeta_{k+3} > 1$, and so

$$(B_{k+1} - Y_k) \parallel (B_{k+1} - Y_k)t + s \parallel < B_{k+1}|D_k|\frac{9+3}{3(9+1)} < \frac{2}{5} < \frac{1}{\sqrt{5}},$$

since $B_{k+1}|D_k| < 1$. On the other hand, if $c_{k+2} > a_{k+2}/3 - 1$, then we consider both $b = B_{k+2} - Y_{k+1}$ and $B_{k+1} + B_{k+2} - Y_{k+1}$. For $\parallel bt + s \parallel$ we have

$$\parallel (B_{k+2} - Y_{k+1})t + s \parallel = |D_{k+2} + R_{k+2}|,$$

$$\parallel (B_{k+1} + B_{k+2} - Y_{k+1})t + s \parallel = |D_{k+1} + D_{k+2} + R_{k+2}|$$

$$= \left|D_{k+1} - |D_{k+2} + R_{k+2}|\right|,$$

since the D_k's alternate signs, and so

$$\left(\parallel (B_{k+2} - Y_{k+1})t + s \parallel\right)\left(\parallel (B_{k+1} + B_{k+2} - Y_{k+1})t + s \parallel\right) < \left(\frac{1}{2}|D_{k+1}|\right)^2 \tag{3}$$

To estimate the b's, we have from (I.1.2) that

$$B_{k+1} = \frac{B_{k+2} - B_k}{a_{k+2}} \tag{4}$$

and also that $B_{k+2} > (a_{k+2} + 1)B_k$, so $B_k/B_{k+2} < 1/(a_{k+2} + 1)$. Now,

$$B_{k+2} - Y_{k+1} < B_{k+2} - c_{k+2}B_{k+1} = B_{k+2}\left(1 - \frac{c_{k+2}}{a_{k+2}} + \frac{B_k}{B_{k+2}}\frac{c_{k+2}}{a_{k+2}}\right) \tag{5}$$

gives

$$B_{k+2} - Y_{k+1} < B_{k+2}\left(1 - \left(\tfrac{1}{3} - \tfrac{1}{9}\right) + \tfrac{1}{10}\left(\tfrac{1}{3} - \tfrac{1}{9}\right)\right) = \tfrac{4}{5}B_{k+2}$$

and (5) combined with (4) gives

$$B_{k+1} + B_{k+2} - Y_{k+1}$$

$$< B_{k+2}\left(1 - \frac{c_{k+2}}{a_{k+2}} + \frac{B_k}{B_{k+2}}\frac{c_{k+2}}{a_{k+2}} + \frac{1}{a_{k+2}} - \frac{B_k}{B_{k+2}}\frac{1}{a_{k+2}}\right) \tag{6}$$

so that

$$B_{k+1} + B_{k+2} - Y_{k+1} < B_{k+2}\left(\tfrac{4}{5} + \tfrac{1}{9} - \tfrac{1}{10}\tfrac{1}{9}\right) = \tfrac{9}{10}B_{k+2}$$

and hence

$$\left(B_{k+2} - Y_{k+1}\right)\left(B_{k+1} + B_{k+2} - Y_{k+1}\right) < \tfrac{18}{25}B_{k+2}^2. \tag{7}$$

Since (3) and (7) mean that at least one of these choices for b satisfies

$$b\,\|\,bt + s\,\| < \tfrac{1}{2}\sqrt{\tfrac{18}{25}}\,B_{k+2}\,|D_{k+1}| < \tfrac{1}{2}\sqrt{\tfrac{18}{25}},$$

we are done with this case.

We may now suppose that t has bounded partial quotients.

Case 2: $a_k \leq 8$ for all k's and $a_{k+2} = 8$ infinitely times. For such an a_{k+2}, if $c_{k+2} \leq 2$, we choose $b = B_{k+1} - Y_k$ and proceed as in (1) and (2) to obtain

$$(B_{k+1} - Y_k)\,\|\,(B_{k+1} - Y_k)t + s\,\| < B_{k+1}\,|D_k|\left(\frac{3\zeta_{k+3} + 1}{8\zeta_{k+3} + 1}\right), \tag{8}$$

which again is a positive, monotone decreasing function of ζ_{k+3}. Since the partial quotients are bounded by 8, the smallest possible value for ζ_{k+3} is $[\overline{1;8}]$ $= (2 + \sqrt{6})/4$ and (8) becomes

$$(B_{k+1} - Y_k) \, \| \, (B_{k+1} - Y_k)t + s \, \| \; < \; \frac{10 + 3\sqrt{6}}{20 + 8\sqrt{6}}.$$

If $c_{k+2} \geq 3$, then proceeding as in (3), (5) and (6) we have

$$B_{k+2} - Y_{k+1} \; < \; B_{k+2}\left(1 - \frac{3}{8} + \frac{1}{9} \cdot \frac{3}{8}\right) = \frac{2}{3} B_{k+2}$$

$$B_{k+1} + B_{k+2} - Y_{k+1} \; < \; B_{k+2}\left(\frac{2}{3} + \frac{1}{8} - \frac{1}{9} \cdot \frac{1}{8}\right) = \frac{7}{9} B_{k+2}$$

and at least one of these choices for b results in

$$b \, \| \, bt + s \, \| \; < \; \frac{1}{2}\sqrt{\frac{14}{27}}.$$

It now suffices to consider only t's with smaller partial quotients.

Case 3: $a_k \leq 7$ for all k's and $a_{k+2} = 7$ infinitely times. For such an a_{k+2}, we proceed as in the previous case. If $c_{k+2} \leq 1$, then (8) becomes

$$(B_{k+1} - Y_k) \, \| \, (B_{k+1} - Y_k)t + s \, \| \; < \; B_{k+1} | D_k | \left(\frac{2\zeta_{k+3} + 1}{7\zeta_{k+3} + 1}\right)$$

with $\zeta_{k+3} = [\overline{1;7}] = (7 + \sqrt{77})/14$ giving the largest possible value, hence

$$(B_{k+1} - Y_k) \, \| \, (B_{k+1} - Y_k)t + s \, \| \; < \; \frac{28 + 2\sqrt{77}}{63 + 7\sqrt{77}}.$$

If $c_{k+2} \geq 2$, then

$$B_{k+2} - Y_{k+1} \; < \; B_{k+2}\left(1 - \frac{2}{7} + \frac{1}{8} \cdot \frac{2}{7}\right) = \frac{3}{4} B_{k+2}$$

$$B_{k+1} + B_{k+2} - Y_{k+1} \; < \; B_{k+2}\left(\frac{3}{4} + \frac{1}{7} - \frac{1}{8} \cdot \frac{1}{7}\right) = \frac{7}{8} B_{k+2}$$

and at least one of these choices for b results in

$$b \, \| \, bt + s \, \| \; < \; \frac{1}{2}\sqrt{\frac{21}{32}}.$$

Case 4: $a_k \leq 6$ for all k's and $a_{k+2} = 6$ infinitely times. Exactly as with Case 3, we have if $c_{k+2} \leq 1$ that

$$(B_{k+1} - Y_k) \, \| \, (B_{k+1} - Y_k)t + s \, \| \; < \; B_{k+1} \, | \, D_k \, | \left(\frac{2\zeta_{k+3} + 1}{6\zeta_{k+3} + 1} \right)$$

with $\zeta_{k+3} = [\overline{1; 6}] = (3 + \sqrt{15})/6$ giving

$$(B_{k+1} - Y_k) \, \| \, (B_{k+1} - Y_k)t + s \, \| \; < \; \frac{6 + \sqrt{15}}{12 + 3\sqrt{15}},$$

while if $c_{k+2} \geq 2$, then

$$B_{k+2} - Y_{k+1} \; < \; B_{k+2}\left(1 - \frac{1}{3} + \frac{1}{7} \, \frac{1}{3} \right) \; = \; \frac{5}{7} B_{k+2}$$

$$B_{k+1} + B_{k+2} - Y_{k+1} \; < \; B_{k+2}\left(\frac{5}{7} + \frac{1}{6} - \frac{1}{7} \, \frac{1}{6} \right) \; = \; \frac{6}{7} B_{k+2}$$

and at least one of these choices for b results in

$$b \, \| \, bt + s \, \| \; < \; \frac{1}{2}\sqrt{\frac{30}{49}}.$$

In each of these four situations, we have been able to obtain an upper bound $< 1/\sqrt{5}$ without considering $B_{k+1} | D_k |$ or $B_{k+2} | D_{k+1} |$ beyond noting that they are less than one. For the remaining cases, we shall also need an estimate for $B_{k+1} | D_k |$. Since the partial quotients are now bounded, let us suppose that $a_k \leq N$ for all k's. Then

$$B_{k+1} | D_k | \; = \; \frac{B_{k+1}}{B_k \zeta_{k+1} + B_{k-1}} \; = \; \frac{B_{k+1}}{B_k\left(a_{k+1} + \dfrac{1}{\zeta_{k+2}} \right) + B_{k-1}}$$

$$= \; \frac{B_{k+1}\zeta_{k+2}}{B_{k+1}\zeta_{k+2} + B_k} \; = \; \frac{\zeta_{k+2}}{\zeta_{k+2} + \xi_{k+1}}$$

is a positive, monotone increasing function of $\zeta_{k+2} > 1$. Thus the maximum occurs when $\zeta_{k+2} = [\overline{N;1}]$ and with ξ_{k+1} as small as possible; that is, $\xi_{k+1} = 1/[\overline{N;1}]$. Hence

$$B_{k+1}|D_k| \leq \frac{[\overline{N;1}]^2}{[\overline{N;1}]^2 + 1} = \frac{2N\big((N+2) + \sqrt{N^2 + 4N}\big)}{2N\big((N+2) + \sqrt{N^2 + 4N}\big) + 4}. \tag{9}$$

We now continue to reduce the size of the partial quotients.

Case 5: $a_k \leq 5$ for all k's and $a_{k+2} = 5$ infinitely times. If $c_{k+2} = 0$, then, as in (8),

$$(B_{k+1} - Y_k)\,\|\,(B_{k+1} - Y_k)t + s\,\| \;<\; B_{k+1}|D_k|\left(\frac{\zeta_{k+3} + 1}{5\zeta_{k+3} + 1}\right)$$

and $\zeta_{k+3} = [\overline{1;5}] = (5 + 3\sqrt{5})/10$ gives the upper bound

$$(B_{k+1} - Y_k)\,\|\,(B_{k+1} - Y_k)t + s\,\| \;<\; \frac{15 + 3\sqrt{5}}{35 + 15\sqrt{5}},$$

while if $c_{k+2} \geq 1$, then

$$B_{k+2} - Y_{k+1} \;<\; B_{k+2}\Big(1 - \frac{1}{5} + \frac{1}{6}\,\frac{1}{5}\Big) \;=\; \frac{5}{6}B_{k+2}$$

$$B_{k+1} + B_{k+2} - Y_{k+1} \;<\; B_{k+2}\Big(\frac{5}{6} + \frac{1}{5} - \frac{1}{6}\,\frac{1}{5}\Big) \;=\; B_{k+2}$$

so at least one of these choices for b results in

$$b\,\|\,bt + s\,\| \;<\; \frac{1}{2}\sqrt{\frac{5}{6}}\;B_{k+2}|D_{k+1}|.$$

Since $(\sqrt{5/6})/2 > 1/\sqrt{5}$, we use (9) with $N = 5$ to obtain

$$b\,\|\,bt + s\,\| \;<\; \frac{35 + 15\sqrt{5}}{37 + 15\sqrt{5}}\,\frac{1}{2}\sqrt{\frac{5}{6}}$$

and this case is finished.

Case 6: $a_k \leq 4$ for all k's and $a_{k+2} = 4$ infinitely times. Repeating the calculations of Case 5 with 4 in place of 5, we have that if $c_{k+2} = 0$, then

$$(B_{k+1} - Y_k)\,\|\,(B_{k+1} - Y_k)t + s\,\| \;<\; B_{k+1}|D_k|\left(\frac{\zeta_{k+3} + 1}{4\zeta_{k+3} + 1}\right)$$

and $\zeta_{k+3} = [\overline{1;\,4}] = (1+\sqrt{2})/2$ gives the upper bound

$$(B_{k+1}-Y_k)\,\|\,(B_{k+1}-Y_k)t+s\,\| \;<\; \frac{3+\sqrt{2}}{6+4\sqrt{2}},$$

while if $c_{k+2} \geq 1$, then

$$B_{k+2}-Y_{k+1} \;<\; B_{k+2}\!\left(1-\tfrac{1}{4}+\tfrac{1}{5}\,\tfrac{1}{4}\right) \;=\; \tfrac{4}{5}B_{k+2}$$

$$B_{k+1}+B_{k+2}-Y_{k+1} \;<\; B_{k+2}$$

so at least one of these choices for b results in

$$b\,\|\,bt+s\,\| \;<\; \frac{[4;\,1]^2}{[4;\,1]^2+1}\,\frac{1}{2}\sqrt{\frac{4}{5}} \;=\; \frac{12+8\sqrt{2}}{13+8\sqrt{2}}\,\frac{1}{\sqrt{5}} \;<\; \frac{1}{\sqrt{5}}.$$

Case 7: $a_k \leq 3$ for all k's and $a_{k+2} = 3$ infinitely times. As in the previous two cases, if $c_{k+2} = 0$, then

$$(B_{k+1}-Y_k)\,\|\,(B_{k+1}-Y_k)t+s\,\| \;<\; B_{k+1}\,|D_k|\!\left(\frac{\zeta_{k+3}+1}{3\zeta_{k+3}+1}\right)$$

and $\zeta_{k+3} = [\overline{1;\,3}] = (3+\sqrt{21})/6$ gives the upper bound

$$(B_{k+1}-Y_k)\,\|\,(B_{k+1}-Y_k)t+s\,\| \;<\; \frac{9+\sqrt{21}}{15+3\sqrt{21}}\,B_{k+1}\,|D_k|.$$

Since $(9+\sqrt{21})/(15+3\sqrt{21}) > 1/\sqrt{5}$, we use (9) with $N = 3$ to obtain

$$(B_{k+1}-Y_k)\,\|\,(B_{k+1}-Y_k)t+s\,\| \;<\; \frac{9+\sqrt{21}}{15+3\sqrt{21}}\,\frac{15+3\sqrt{21}}{17+3\sqrt{21}} \;<\; \frac{1}{\sqrt{5}}.$$

For $c_{k+2} \geq 1$, we have

$$B_{k+2}-Y_{k+1} \;<\; B_{k+2}\!\left(1-\tfrac{1}{3}+\tfrac{1}{4}\,\tfrac{1}{3}\right) \;=\; \tfrac{3}{4}B_{k+2}$$

and $B_{k+1}+B_{k+2}-Y_{k+1} < B_{k+2}$, so at least one of these choices for b gives

$$b \parallel bt + s \parallel \; < \; \frac{1}{2}\sqrt{\frac{3}{4}}\, B_{k+2} \, | \, D_{k+1} \, |$$

and we are done with this case.

Case 8: $a_k \leq 2$ for all k's with equality infinitely many times. If we had $a_{k+2} = 2$ and $c_{k+2} = 0$, the calculations of Case 7 with 2 in place of 3 would give only that

$$(B_{k+1} - Y_k) \parallel (B_{k+1} - Y_k)t + s \parallel \; < \; \frac{3 + 3\sqrt{3}}{5 + 2\sqrt{3}}.$$

Instead, let us first suppose that $c_{k+2} = a_{k+2}$ infinitely many times. Then $c_{k+1} = 0$, $c_{k+3} < a_{k+3}$,

$$\parallel (B_k - Y_{k-1})t + s \parallel \; = \; | \, D_k + 0 \cdot D_k + a_{k+2} D_{k+1} + c_{k+3} D_{k+2} + \cdots \, | \; < \; | \, D_{k+1} \, |$$

and so

$$(B_k - Y_{k-1}) \parallel (B_k - Y_{k-1})t + s \parallel \; < \; B_k \, | \, D_{k-1} \, | \frac{1}{a_{k+1}\zeta_{k+2} + 1}$$

$$< \; \frac{1}{[\overline{1;2}] + 1} \; = \; \frac{2}{3 + \sqrt{3}} \; < \; \frac{1}{\sqrt{5}}.$$

We now may suppose that $c_k \leq a_k - 1$ for all k's and that $a_{k+2} = 2$ infinitely many times. If $c_{k+2} = 0$ for such an a_{k+2}, then

$$(B_{k+1} - Y_k) \parallel (B_{k+1} - Y_k)t + s \parallel \; < \; B_{k+1} \, | \, D_{k+1} + 0 \cdot D_{k+1} + c_{k+3} D_{k+2} + \cdots \, |$$

$$\leq \; B_{k+1} \, | \, D_{k+1} + (a_{k+4} - 1)D_{k+3} + (a_{k+6} - 1)D_{k+5} + \cdots \, | .$$

Since $a_k - 1 \leq a_k/2$, we find that this expression is

$$\leq \; B_{k+1} \left| D_{k+1} - \frac{1}{2} D_{k+2} \right| \; = \; B_{k+1} \, | \, D_k \, | \frac{\zeta_{k+3} + \frac{1}{2}}{a_{k+3}\zeta_{k+3} + 1} \; = \; \frac{1}{2} B_{k+1} \, | \, D_k \, |$$

and, by taking $N = 2$ in (9), we have

$$(B_{k+1} - Y_k) \parallel (B_{k+1} - Y_k)t + s \parallel \; < \; \frac{2 + \sqrt{3}}{5 + 2\sqrt{3}} \; < \; \frac{1}{\sqrt{5}}.$$

If $c_{k+2} = 1$, we have

$$B_{k+2} - Y_{k+1} < B_{k+2}\left(1 - \tfrac{1}{2} + \tfrac{1}{3}\tfrac{1}{2}\right) = \tfrac{2}{3}B_{k+2}$$

and $B_{k+1} + B_{k+2} - Y_{k+1} < B_{k+2}$, so at least one of these choices for b gives

$$b \parallel bt + s \parallel \ < \ \tfrac{1}{2}\sqrt{\tfrac{2}{3}}\, B_{k+2} |D_{k+1}|$$

and this case if finished.

Case 9: $t \sim g$. If $c_{k+2} = 0$ for all sufficiently large k's, then the result follows from Theorem 5.3. If the c_{k+2}'s consist of alternating 0's and 1's from some point on, by the discussion leading up to Theorem II.6.1, s is of the form $mt + n$ where m and n are integers and $b \parallel bt + s \parallel$ again reduces to the homogeneous case. Otherwise, we must have $c_{k+1} = c_{k+2} = 0$ and $c_{k+3} = 1$ infinitely many times. Since $c_{k+3} = 1$ and $a_{k+4} = 1$, we must have $c_{k+4} = 0$. Choosing $b = B_{k+1} - Y_k$, we have

$$B_{k+1} - Y_k \ = \ B_{k+1} - Y_{k-1} < B_{k+1} - B_{k-1} \ = \ B_k$$

and

$$\parallel (B_{k+1} - Y_k)g + s \parallel \ = \ |D_{k+1} + D_{k+2} + 0 \cdot D_{k+3} + \cdots |$$

$$= \ |D_{k+3} + \cdots | \ < \ |D_{k+3} - D_{k+4}|,$$

since $D_{k+1} + D_{k+2} = D_{k+3}$. Finally, since $\zeta_k = g$, we have

$$(B_{k+1} - Y_k) \parallel (B_{k+1} - Y_k)g + s \parallel \ < \ B_k |D_{k+3}|\left(1 + \tfrac{1}{g}\right) < \tfrac{1}{g^3}$$

and the proof is completed.

§10. Szekeres' empty parallelogram theorem.

Let α and β be two different irrational numbers and consider the parallelogram in the plane with sides parallel to the lines $y = \alpha x$ and $y = \beta x$ and with center at the origin (see Figure 1). It is clear that such a parallelogram can be drawn so that it contains no lattice points other than the origin (a *lattice point* is a point with integer coordinates). Minkowski considered the upper bound Δ of the areas of such figures and his classical theorem on the geometry of numbers states that $\Delta \leq 4$, independent of α and β. Mordell showed that $\Delta \geq 2$. We shall prove Szekeres' result that the best possible value of Δ is $4g/\sqrt{5}$ in the sense that $\Delta \geq 4g/\sqrt{5}$ for any α and β and that there are α and β such that any parallelogram with area greater than $4g/\sqrt{5}$ must contain interior lattice points (here $g = (1 + \sqrt{5})/2$ is the "golden ratio" described in §7 of Chapter I).

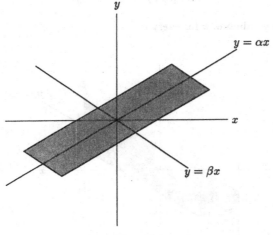

Figure 1.

Let P denote the parallelogram of Figure 1, which we now suppose to be free of interior lattice points except the origin. Since P is symmetric, the x-coordinates must be $|x| < b$ or even $|x| \leq b$ for some smallest positive integer b. Let us expand α as a continued fraction with convergents $\{A_k/B_k\}_{k \geq -1}$ and suppose that $B_k < b \leq B_{k+1}$. From Theorem II.3.1, $|X\alpha - Y| \geq |B_k\alpha - A_k|$ for any integers X and Y with $|X| \leq b$. Thus the vertical thickness of P can be no more than twice $|B_k\alpha - A_k|$. Since the parallelogram P* given by

$$\begin{cases} -B_{k+1} < x < B_{k+1} \\ \alpha x - |B_k\alpha - A_k| < y < \alpha x + |B_k\alpha - A_k| \end{cases}$$

(see Figure 2) has the largest area of any such parallelogram (and, in fact, has lattice points on its boundary), it suffices to consider only parallelograms P* for large values of k. The area of P* is

$$4B_{k+1} \, | \, B_k\alpha - A_k| \;=\; \frac{4B_{k+1}\zeta_{k+2}}{B_{k+1}\zeta_{k+2} + B_k}.$$

We shall prove that

$$\frac{B_{k+1}\zeta_{k+2}}{B_{k+1}\zeta_{k+2} + B_k} \;\geq\; \frac{g}{\sqrt{5}} \qquad\qquad (1)$$

for infinitely many values of k for every α.

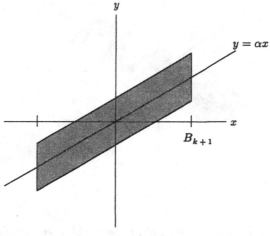

Figure 2.

First, suppose that $\zeta_{k+2} > 3$ for infinitely many k's. Then

$$\frac{B_{k+1}\zeta_{k+2}}{B_{k+1}\zeta_{k+2} + B_k} \;>\; \frac{3}{4} \;>\; \frac{g}{\sqrt{5}},$$

since $x/(x + B_k) \uparrow 1$ as $x \to \infty$ and $B_k/B_{k+1} < 1$, and we are done.

Second, suppose that $\zeta_{k+2} > 2$ and

$$\frac{B_{k+1}\zeta_{k+2}}{B_{k+1}\zeta_{k+2}+B_k} < \frac{g}{\sqrt{5}}$$

for infinitely many k's. We shall show that we must have had $\zeta_k > 3$ and then the claim will follow from the previous case. Given the conditions,

$$\frac{2B_{k+1}}{2B_{k+1}+B_k} < \frac{g}{\sqrt{5}}$$

and so

$$B_{k+1} < \frac{1+g}{2}\, B_k.$$

Since $B_{k+1} = a_{k+1}B_k + B_{k-1}$ and $(1+g)/2 < 2$,

$$a_{k+1} + \xi_k < 2$$

and thus $a_{k+1} = 1$. But now

$$\xi_k < \frac{1+g}{2} - 1 < \frac{1}{3},$$

so $a_k \ge 3$ and hence $\zeta_k > 3$.

Finally, we must consider the possibility that $\zeta_{k+2} < 2$ for all sufficiently large values of k. But then $a_{k+2} = 1$ for k large and so $\zeta_{k+2} = g$. Thus $\xi_k \to 1/g$ as $k \to \infty$ and

$$\lim_{k\to\infty} \frac{B_{k+1}\zeta_{k+2}}{B_{k+1}\zeta_{k+2}+B_k} = \lim_{k\to\infty} \frac{\zeta_{k+2}}{\zeta_{k+2}+\xi_{k+1}} = \frac{g}{g+\frac{1}{g}} = \frac{g}{\sqrt{5}}.$$

Since ξ_k approximates $1/g$ alternately from above and below, there are infinitely many k's for which $\xi_k < 1/g$, which means that there are infinitely many k's which satisfy (1) and we are finished.

It is clear that there is no improvement possible for the kind of α's considered in the final case. However, if the ζ_k's are unbounded then $B_{k+1}\zeta_{k+2}/(B_{k+1}\zeta_{k+2}+B_k)$ can be arbitrarily close to 1 from below and hence Minkowski's estimate that $\Delta \le 4$ is the best possible upper bound. In Chapter V we shall show that "most" numbers have unbounded complete quotients.

NOTES

§1. See Huygens [(1681), (1682), 1703]; sketches of the gears appear on pages 608 and 612 and photographs appear as "Fig. 61" between pages 180 and 181. Huygens also considered the continued fraction of π (see Huygens [1703], pages 632-635 and [(1686)]). His planetarium was constructed by the clockmaker Johannes van Ceulen and, according to Cousins [1972], is "... to be seen today in the Rijksmuseum Voor de Geschiedenis der Natuurwentenschappen at Leyden." Weil [1984] notes that Lagrange later gave full credit to Huygens for his study of continued fractions and best approximations.

The Gregorian revision of the Julian calendar also addressed the calculation of Easter, which requires estimates of the relation between the tropical year and the synodic month. There is no evidence that continued fractions were used in the construction of the Gregorian calendar; however, many subsequent writers have attempted (unsuccessfully) to use continued fractions to justify $365\frac{97}{400}$ as an approximation to their value of the tropical year (see Dutka [1988]).

Other methods to solve gear ratio problems have been proposed. "Piper's algorithm," described in Mathews [1990], factors the numerators and denominators of fractions "close" to the desired value and then attempts to regroup these factors to form two terms within the required range. Hua (see Halberstam [1986]) gave a solution for the value π by using Farey sequences to decompose the numerators and denominators of possible gear trains in terms of the convergents of π; however, he did not rediscover the Ostrowski representation (for example, he wrote $2465 = 20 \cdot 113 + 29 \cdot 7 + 2 \cdot 1$) and "... decided that this technique was not sufficiently general for popularisation." We must remark that, while not aesthetically appealing, the obvious "brute force" solution may be carried out easily on a personal computer.

§2. See Dickson [1971], II, Chapter XII, where a long history may be found, including the connection with the "cattle problem of Archimedes." Dickson, page 341, remarks that $x^2 - dy^2 = 1$ "... should have been designated as Fermat's equation" while LeVeque [1977], page 205, notes that "... probably the name sticks because it is unambiguous, since Pell also did little else of any great mathematical merit." The results for $|N| > \sqrt{d}$ are from Rockett and Szüsz [1986].

§3. The four known constructions are due to Legendre (1808), Gauss (1825), Hermite/Serret (1848) and Jacobsthal (1906). Serret's construction depends on the continued fraction for p/x while Hermite's uses that for x/p (see also Brillhart [1972]). Dickson [1971], II, Chapter VI, and Davenport [1982], Chapter V, describe all four constructions and their "computational complexity" is discussed by Shanks [1967]. It suffices to represent primes as sums of two squares since the classical identity $(a^2 + b^2)(c^2 + d^2) = (ac + bd)^2 + (ad - bc)^2$ extends the representations to products. The number of possible representations is discussed in LeVeque [1977], Chapter 7, and Niven and Zuckerman [1980], Chapter 5. Our presentation of Legendre's construction follows Legendre [1830], pages 64-71.

§4. Our presentation is based on Hall [1947]; see also Cusick and Flahive [1989], Chapter 4, where additional refinements will be found. The non-constructive demonstration that $\mathcal{C} + \mathcal{C} = [0, 2]$ is intended only as an orientation for the proof of Theorem 1. A direct construction for $t = t_1 + t_2$ is simple to give. Letting $d_k(x)$ be the k-th digit in the ternary expansion of x, if $d_k(t) \neq 1$ for $k \geq 0$ then we can just take $d_k(t_1) = 0$ and $d_k(t_2) = d_k(t)$ for $k \geq 0$. If $d_0(t) = 1$ while $d_k(t) \neq 1$ for $k \geq 1$, we can set $d_0(t_1) = d_0(t_2) = 0$ and then take $d_k(t_1) = 2$ and $d_k(t_2) = d_k(t)$ for $k \geq 1$. For any t in $[0, 2]$, the digits that are 1's can be viewed as "switches" that signal a change between $d_k(t_1)$ being 0 or 2: when a first $d_k(t) = 1$ activates the switch, we set $d_k(t_1) = d_k(t_2) = 0$ and then continue for subsequent k's by setting $d_k(t_1) = 2$ and $d_k(t_2) = d_k(t)$ until we reach the next $d_k(t) = 1$, which turns the switch off with $d_k(t_1) = d_k(t_2) = 2$ (the construction now continues as before with $d_k(t_1) = 0$ and $d_k(t_2) = d_k(t)$ until another $d_k(t) = 1$ is encountered). For example, $1 = 0.\overline{2} + 0$, $0.\overline{1} = 0.\overline{02} + 0.\overline{02}$, and $1.2022010222101 = 0.2222220000022 + 0.2022020222002$.

§5. See Perron [1954], §14. Theorem 1 is due to Vahlen [1895]. The method of proof for Theorem 2 originates with Borel [1903]. Although $1/\sqrt{5}$ was already known as the bound for indefinite, binary quadratic forms (see §6), Theorem 2 is known as "Hurwitz's theorem" because he gave it in the form of a statement about diophantine approximation. Theorem 4 is a refinement by Wright [1964] of an observation of Forder [1963].

§6. Our presentation is a hybrid of continued fractions (see Dickson [1930],

Chapter VII, and also LeVeque [1977]) and quadratic forms (see Dickson [1929], Chapters VII and XI, Remak [1924] and Cassels [1972], Chapter II); see also Weil [1984], §IV of Chapter IV, Appendix II ("A proof of Legendre's on positive binary quadratic forms") and Appendix III ("A proof of Lagrange's on indefinite binary quadratic forms"). That $\mathfrak{M}$ contains numbers other than those of $\mathfrak{L}$ was shown in 1968 by Freiman. The reciprocal of the interval described in Theorem 2 forms "Hall's ray" and the lower endpoint for Hall's ray was found by Freiman in 1975. See Cusick and Flahive [1989] for these and other results; see also Series [1985].

§7. See Szüsz [1973], Mahler [1945], and also LeVeque [1953-54], where Farey sequences are used to the same end. In the special case when $K = 1/n$, where n is a positive integer, it is not possible to improve the estimates of the theorem. However, if $K = 1/(n+c)$, where $0 < c < 1$, then it is possible to reduce the positive upper estimate.

§9. Our proofs using the t-expansion present a uniform method for these inhomogeneous problems. See also Barnes [1954], Barnes and Swinnerton-Dyer [1952a,b; 1954], Cassels [1972], Chapter III, Davenport [1947, 1950], Khintchine [1926a] and Koksma [1936], Kapitel VI. As might be anticipated from the proof of Theorem 4, it is possible to carry out a study of $b \parallel bt + s \parallel$ below $1/\sqrt{5}$ analogous to the Markov spectrum for $b \parallel bt \parallel$ between $1/\sqrt{5}$ and $1/3$. The closeness of the estimate in Case 6 gives a clue to the next class of interesting t's after $t \sim g$; see Descombes [1956a,b,c].

§10. See Hardy and Wright [1971], Szekeres [1937] and Szüsz [1956].

Chapter V

METRICAL THEORY

We now shift our point of view and investigate statements which are true for "most" but not all numbers. We saw in the previous chapter that our results were determined by numbers with bounded partial quotients; stronger statements would have been possible if the partial quotients had been unbounded. We begin our study of the metrical theory of continued fractions by investigating the "probability" that a number chosen at random has bounded partial quotients. Since the initial partial quotient a_0 is irrelevant to our discussion, we shall restrict our attention for the rest of this book to numbers t in the unit interval $X = [0, 1)$.

A class $\mathscr{S}$ of subsets of X is a *σ-algebra* if (a) $X \in \mathscr{S}$, (b) $S \cap T^c \in \mathscr{S}$ whenever S, $T \in \mathscr{S}$ (here T^c denotes the *complement* of T in X so that $T^c \cap T = \emptyset$ and $T^c \cup T = X$), and (c)

$$\bigcup_{k=1}^{\infty} S_k \in \mathscr{S}$$

whenever $S_k \in \mathscr{S}$ for $k = 1, 2, \ldots$. We shall be interested in the *Borel sets* of X, which are the elements of the smallest σ-algebra that contains the intervals $[x, y)$, where $0 \le x < y < 1$. We shall refer to these Borel sets both as "sets" and as "events" and to each such set S we associate a *probability* $P(S)$, which is the Lebesgue measure of the set. Since the Lebesgue measure of an interval is its length, we shall also write $|S|$ for $P(S)$ if S is an interval. It follows that $P(X) = 1$, $P(S) \ge 0$ for any S, and

$$P\left(\bigcup_{k=1}^{\infty} S_k\right) = \sum_{k=1}^{\infty} P(S_k)$$

if the S_k are *pairwise disjoint* (that is, $S_i \cap S_j = \emptyset$ for $i \ne j$). By a set of *measure zero* or an event with *probability zero* we mean a set S with $P(S) = 0$, while a set that is *almost everywhere* or an event that occurs with *probability one* is a set S with $P(S) = 1$. A set S of measure zero is such that given any $\epsilon > 0$ there is a cover of S consisting of intervals T_k ($k = 1, 2, \ldots$) such that every element of S is contained in at least one T_k and $\sum P(T_k) < \epsilon$. Thus every countable set of numbers is of measure zero, since T_k may be taken to be the interval of length $\epsilon/2^k$ centered about the k-th number.

§1. Numbers with bounded partial quotients.

An *interval of order* n, $I_n = I_n(k_1, \ldots, k_n)$, consists of those numbers $t \in [0,1)$ such that $a_1(t) = k_1$, $a_2(t) = k_2$, $\ldots$, $a_n(t) = k_n$. Thus every $t \in I_n$ is of the form $(A_n\zeta_{n+1} + A_{n-1})/(B_n\zeta_{n+1} + B_{n-1})$ with the same A_n, A_{n-1}, B_n and B_{n-1} where, of course, $\zeta_{n+1} \geq 1$ and so

$$|I_n| = \left| \frac{A_n + A_{n-1}}{B_n + B_{n-1}} - \frac{A_n}{B_n} \right| = \frac{1}{B_n(B_n + B_{n-1})} = \frac{1}{B_n^2(1 + \xi_n)}. \tag{1}$$

Since each $t \in [0,1)$ is in exactly one interval of order n, we see that

$$\bigcup_{k_1, \ldots, k_n} I_n(k_1, \ldots, k_n) = [0,1), \tag{2}$$

where the union runs over all possible n-tuples of values $(k_1, \ldots, k_n)$. We shall write $I_{n+1}(k)$ for the interval of order $n+1$ formed from I_n by the additional requirement that $a_{n+1}(t) = k$; that is, $t \in I_{n+1}(k)$ means that $t = (A_n\zeta_{n+1} + A_{n-1})/(B_n\zeta_{n+1} + B_{n-1})$ where $k \leq \zeta_{n+1} < k+1$.

Theorem 1. *The set of numbers in* $[0,1)$ *with bounded partial quotients has measure zero.*

Proof. Let $N > 1$ be given and let us write

$$S_N = \{t \in [0,1) : a_k < N \text{ for } k = 1, 2, \ldots\}$$

and

$$S_N(n) = \{t \in [0,1) : a_k < N \text{ for } k = 1, 2, \ldots, n\},$$

so that $S_N \subset S_N(n)$ and each $S_N(n)$ is a finite union of disjoint intervals of order n, $I_n = I_n(k_1, \ldots, k_n)$ where $k_i < N$ for $i = 1, \ldots, n$. For a given $I_n \subset S_N(n)$, all the intervals of order $n+1$ contained in $S_N(n+1)$ and in I_n are of the form $I_{n+1}(k)$ with $1 \leq k < N$. But

$$\left| \bigcup_{1 \leq k < N} I_{n+1}(k) \right| = \left| \frac{A_n + A_{n-1}}{B_n + B_{n-1}} - \frac{A_n N + A_{n-1}}{B_n N + B_{n-1}} \right|$$

$$= \frac{1}{B_n^2(1+\xi_n)} \cdot \frac{N-1}{N+\xi_n} < (1 - \frac{1}{N}) \cdot |I_n|. \tag{3}$$

Thus

$$P(S_N(n+1)) < (1 - \frac{1}{N}) \cdot P(S_N(n)) < \cdots < (1 - \frac{1}{N})^n \cdot P(S_N(1))$$

and so $P(S_N(n)) \to 0$ as $n \to \infty$. Thus for any $N > 1$, $P(S_N) = 0$ and the theorem follows.

We now show that if the partial quotients are bounded by a function that grows "slowly" (or not at all as in Theorem 1), then the measure of such numbers is still zero.

Theorem 2. *Let $f(n) > 1$ for $n = 1, 2, \ldots$ and suppose that $\sum 1/f(n)$ diverges. Then the set*

$$S = \{t \in [0,1) : a_n < f(n) \text{ for } n = 1, 2, \ldots\}$$

has measure zero.

Proof. From (3), we see that

$$\left| \bigcup_{1 \leq k < f(n+1)} I_n(k) \right| < (1 - \frac{1}{f(n+1)}) \cdot |I_n|.$$

Writing $S(n) = \{t \in [0,1) : a_1 < f(1), \ldots, a_n < f(n)\}$, we have

$$P(S(n+1)) < (1 - \frac{1}{f(n+1)}) \cdot P(S(n)) < \cdots < P(S(1)) \prod_{k=1}^{n} \left(1 - \frac{1}{f(k+1)}\right).$$

Since $1 - x < \exp(-x)$,

$$\prod_{k=1}^{n}\left(1 - \frac{1}{f(k+1)}\right) < \exp\left(-\sum_{k=1}^{n}\frac{1}{f(k+1)}\right) \tag{4}$$

and so $P(S(n)) \to 0$ as $n \to \infty$. Since $S \subset S(n)$, $P(S) = 0$.

§2. The Borel-Cantelli lemma.

Theorems 1.1 and 1.2 put conditions on all of the partial quotients. As immediate corollaries, we also have that almost all numbers have partial quotients which can be arbitrarily large. But how many numbers have very large partial quotients very often? To deal with this question, we now investigate the probability of an event occurring "infinitely many times" in a given sequence of events.

Theorem 1. *Let* S_1, S_2, ... *be a sequence of events such that* $\sum P(S_k) < \infty$ *and suppose that the event* S *has* $S \subset S_k$ *for infinitely many* k's. *Then* $P(S) = 0$.

Proof. The result is immediate, since given any $\epsilon > 0$ there is an integer N with

$$\sum_{k=N}^{\infty} P(S_k) < \epsilon$$

and we also have that $P(S) < \sum P(S_k)$.

It also is possible to show that the divergence of $\sum P(S_k)$ implies that $P(S) = 1$, provided that the events S_k are "sufficiently independent" in the sense that $P(S_i \cap S_j)$ is essentially $P(S_i) \cdot P(S_j)$ for $i \neq j$. These two results together are known as the *Borel-Cantelli lemma*. As we shall use just Theorem 1 in our subsequent proofs, we shall refer to it as the Borel-Cantelli lemma.

We now show a companion result to Theorem 1.2.

Theorem 2. *Let* $f(n) > 1$ *for* $n = 1, 2, \ldots$ *and suppose that* $\sum 1/f(n) < \infty$. *Then the set*

$$S = \{t \in [0,1) : a_k(t) > f(k) \ \textit{infinitely many times}\}$$

has measure zero.

Proof. Let $S_n = \{t \in [0,1) : a_n > f(n)\}$. As before,

$$\left| \bigcup_{k \geq f(n+1)} I_{n+1}(k) \right| = \left| \frac{A_n f(n+1) + A_{n-1}}{B_n f(n+1) + B_{n-1}} - \frac{A_n}{B_n} \right|$$

$$= \frac{1}{B_n^2(1+\xi_n)} \cdot \frac{1+\xi_n}{f(n+1)+\xi_n} < \frac{2}{f(n+1)} \cdot |I_n|.$$

and so $P(S_{n+1}) < 2/f(n+1)$ by (1.2). The result now follows from Theorem 1.

Thus we see that "most" numbers have partial quotients that, although not bounded, are not too large too often. Before studying the possibility of an "average value" for the partial quotients, we must develop some basic concepts from probability theory.

§3. Random variables and expectations.

A *random variable* f is a real valued function on the probability space X such that for each real number x the set $\{t \in X : f(t) < x\}$ is a Borel set. If f is integrable over the space X in the sense of Lebesgue, then the *expectation* $E(f)$ of f is

$$E(f) = \int f \, dP.$$

We shall use repeatedly two basic properties of the Lebesgue integral without explicit further mention. First, the *Lebesgue Dominated Convergence Theorem*: If g and f_1, f_2, ... are random variables with expectations $E(g)$, $E(f_1)$, $E(f_2)$, ... such that $|f_k| < g$ for $k = 1$, 2, ... and the limit as $k \to \infty$ of f_k exists almost everywhere, then $\lim E(f_k) = E(\lim f_k)$. Second, the *Beppo-Levi Theorem*: If f_1, f_2, ... are non-negative random variables for which $\sum E(f_k) < \infty$, then $\sum f_k$ is convergent almost everywhere and $E(\sum f_k) = \sum E(f_k)$.

Is there an expected value for the partial quotients? Since the partial quotients take positive integer values, the integral in question is of a step function and, consequently, is the same as

$$\sum_{k=1}^{\infty} k \cdot P(a_n = k), \tag{1}$$

where $P(a_n = k)$ denotes the measure of the set of $t \in [0,1)$ with $a_n(t) = k$. From (1.1), in any given interval of order n the proportion of numbers t with $a_{n+1}(t) = k$ is the ratio

$$\frac{\left| \dfrac{A_n(k) + A_{n-1}}{B_n(k) + B_{n-1}} - \dfrac{A_n(k+1) + A_{n-1}}{B_n(k+1) + B_{n-1}} \right|}{\left| \dfrac{A_n + A_{n-1}}{B_n + B_{n-1}} - \dfrac{A_n}{B_n} \right|}$$

$$= \frac{B_n(B_n + B_{n-1})}{(B_n(k) + B_{n-1})(B_n(k+1) + B_{n-1})} = \frac{1 + \xi_n}{k + \xi_n} \cdot \frac{1}{k + (1 + \xi_n)}.$$

Thus this proportion is greater than $(1/3)/(k(k+1))$ and less than $2/(k(k+1))$. Summing over all intervals of order n,

$$\frac{1/3}{k(k+1)} < P(a_n = k) < \frac{2}{k(k+1)}, \tag{2}$$

independently of n. Thus the series (1) is divergent and the expectation of a_n does not exist.

Further, we remark that a limiting value for the arithmetic means

$$\frac{1}{n} \sum_{k=1}^{n} a_k$$

does not exist for almost all numbers. Since $\sum 1/(k \log(k))$ is divergent, Theorem 1.2 shows that the measure of those numbers which always have $a_k < k \log(k)$ is zero, and thus

$$\sum_{k=1}^{n} a_k > n \log(n)$$

infinitely often almost everywhere.

From the estimate of $P(a_n = k)$ in (2), we see that in order for a random variable f to have an expected value on the a_k's, f must increase (if at all) slower than a linear function. We shall return to this observation during the proof of Khintchine's theorem, after we develop some further tools from probability theory and make a more precise estimate of $P(a_n = k)$.

§4. Chebyshev's inequality and large number laws.

Given a random variable f, we write $P(f < x)$ for the probability of the event that $t \in X$ has $f(t) < x$. The function $F(x) = P(f < x)$ is called the *distribution function* of f. Such a function has the following properties: (a) $F(x)$ is non-decreasing; (b) $F(-\infty) = 0$, where $F(-\infty)$ is the limit of $F(x)$ as $x \to -\infty$, and $F(\infty) = 1$, where $F(\infty)$ is limit of $F(x)$ as $x \to \infty$; and (c) $F(x)$ is semi-continuous from the left. $F(x)$ determines a measure μ in the sense that we may define $\mu([x,y)) = F(y) - F(x)$ for intervals $[x,y)$ and then extend this measure to all the Borel sets. The Lebesgue integral of a function $g(x)$ with respect to this measure is the *Lebesgue-Stieltjes integral of g* and we shall write

$$\int g(x)\, d\mu \;=\; \int g\, dF.$$

If μ is absolutely continuous with respect to the usual Lebesgue measure (that is, if $F(x)$ is absolutely continuous), then the derivative $F'(x)$ of $F(x)$ is the *density function* of f and we have that

$$\int g\, dF \;=\; \int g(x) \cdot F'(x)\, dx$$

for any g for which the integral exists. In particular, the *n-th moment* of f is

$$E(f^n) \;=\; \int x^n\, dF$$

and the expectation is just the first moment:

$$E(f) \;=\; \int x\, dF.$$

The *variance* of f is $D^2(f) = E((f - E(f))^2) = E(f^2) - E^2(f)$. We write D for the non-negative square root of the variance.

If f_1, f_2, ... are random variables, then the function $F_n(x_1, \cdots, x_n)$ $= P(f_1 < x_1, \cdots, f_n < x_n)$ is the *joint distribution function* of $f_1, \cdots, f_n$. The random variables are *pairwise independent* if $P(f_i < x_i, f_j < x_j) = P(f_i < x_i)$ $\cdot P(f_j < x_j)$ for any $i \neq j$ and are *totally independent* if $P(f_i < x_i, \cdots, f_j < x_j)$ $= P(f_i < x_i) \cdot \cdots \cdot P(f_j < x_j)$ for any n-tuples of f_i's and x_i's for every n.

If f_1, f_2, ... are pairwise independent random variables with expectations $E(f_k)$ and variances $D^2(f_k)$, then

$$E(f_i f_j) = E(f_i) \cdot E(f_j) \quad \text{for } i \neq j,$$

since the joint distribution is the same as $P(f_i < x_i) \cdot P(f_j < x_j)$ and the double integral may be broken apart. Further,

$$D^2(f_i + f_j) = D^2(f_i) + D^2(f_j) \quad \text{for } i \neq j,$$

since

$$E\Big(\big((f_i + f_j) - E(f_i + f_j)\big)^2\Big) = E\Big(\big((f_i - E(f_i)) + (f_j - E(f_j))\big)^2\Big)$$

$$= E\Big(\big(f_i - E(f_i)\big)^2\Big) + E\Big(\big(f_j - E(f_j)\big)^2\Big) + 2E\Big(\big(f_i - E(f_i)\big) \cdot \big(f_j - E(f_j)\big)\Big),$$

and the last term of this expression is twice

$$E(f_i f_j) - 2E(f_i)E(f_j) + E(f_i)E(f_j),$$

which is zero. It now follows by induction that

$$D^2\Big(\sum_{k=1}^{n} f_k\Big) = \sum_{k=1}^{n} D^2(f_k)$$

for pairwise independent random variables.

We shall show later that the partial quotients of t are *weakly dependent* random variables in the sense that given positive integers n, k, r and s,

$$P(a_n = r \text{ and } a_{n+k} = s) \;=\; P(a_n = r) \cdot P(a_{n+k} = s) \cdot \big(1 + \mathcal{O}(q^k)\big),$$

where the constant q is $0 < q < 1$. Thus for large k's the quantities of interest to us behave like pairwise independent random variables. For this reason, we now study the limiting behavior of sequences of pairwise independent random variables. In addition to the Borel-Cantelli lemma (Theorem 2.1), we shall need an estimate of the probability that the random variable f is far from its expectation $E(f)$ known as *Chebyshev's inequality*.

Theorem 1. *Let f be a random variable with expectation E and variance D^2. Then*

$$P(\,|f - E| > \lambda D) \;\le\; \tfrac{1}{\lambda^2} \tag{1}$$

for any $\lambda > 1$.

Proof. Since $D^2 = \int (x - E)^2 \, dF$, we may separate the integral into two parts:

$$D^2 \;=\; \int\limits_{(x-E)^2 > (\lambda D)^2} (x-E)^2 \, dF \;+\; \int\limits_{(x-E)^2 \le (\lambda D)^2} (x-E)^2 \, dF$$

and so

$$D^2 \;>\; \lambda^2 D^2 \int\limits_{|x-E| > \lambda D} dF,$$

which is the same as (1).

A *law of large numbers* for a sequence $f_1,\ f_2,\ \ldots$ of random variables is the claim that the arithmetic means of the f_k's tend to the limiting arithmetic mean of the expectations. If the $f_1,\ f_2,\ \ldots$ are pairwise independent and identically distributed with expectations $E = E(f_k)$ and variances $D^2 = D^2(f_k)$, then $E(f_1 + \cdots + f_n) = nE$ and $D^2(f_1 + \cdots + f_n) = nD^2$, and we may apply Chebyshev's inequality with $\lambda = \epsilon\sqrt{n}/D$, where $\epsilon > 0$, to find that

$$P\!\left(\left|\sum_{k=1}^{n} f_k - nE\right| > \epsilon n\right) \;<\; \frac{D^2}{n\epsilon^2}$$

and so

$$\lim_{n\to\infty} P\left(\left| \frac{1}{n} \sum_{k=1}^{n} f_k - E \right| > \epsilon \right) = 0 \qquad (2)$$

for any $\epsilon > 0$. A relation of the form (2) is called a *weak law of large numbers* because given an ϵ-neighborhood about E, while the probability that $(1/n)\sum f_k$ is outside this neighborhood goes to zero, no such restriction is claimed for the values of $(1/n)\sum f_k$. In fact, the values of $(1/n)\sum f_k$ may strongly deviate from E infinitely many times while n increases. A famous theorem of Markov extends (2) to pairwise independent random variables f_1, f_2, ... such that $E(f_k)$ and $D^2(f_k)$ exist for $k = 1, 2, \dots$, the limit as $n \to \infty$ of $(1/n)\sum E(f_k)$ is finite, and the limit of $(1/n^2)\sum D^2(f_k)$ is zero. Furthermore, the existence of the variances is not necessary, as we now show by proving a theorem of Khintchine using the important technique of *truncation*.

Theorem 2. *Let f_1, f_2, ... be pairwise independent and identically distributed random variables with expectations $E = E(f_k)$. Then*

$$\lim_{n\to\infty} P\left(\left| \frac{1}{n} \sum_{k=1}^{n} f_k - E \right| > \epsilon \right) = 0$$

for any $\epsilon > 0$.

Proof. We may suppose, without loss of generality, that $E = 0$. Since the expectations $E(f_k)$ exist, so do $E(|f_k|)$ and we shall denote their common value by M. Let $F(x)$ be the common distribution function of the f_k's and let $\epsilon > 0$ be given. We must show that for any $\delta > 0$, there is an integer $N = N(\delta)$ such that

$$P\left(\left| \frac{1}{n} \sum_{k=1}^{n} f_k \right| > \epsilon \right) < \delta \quad \text{for } n > N.$$

Given $\delta > 0$, let $C = \delta\epsilon^2/8M$ and we truncate f_k to form a new random variable f_k^* by setting

$$f_k^* = \begin{cases} f_k & \text{if } |f_k| < nC \\ 0 & \text{otherwise} \end{cases} \qquad (3)$$

for $k = 1, \dots, n$. Let $E^* = E(f_k^*)$. Since

$$P\left(\left|\frac{1}{n}\sum_{k=1}^{n}f_k\right|>\epsilon\right) = P\left(\left|\left(\frac{1}{n}\sum_{k=1}^{n}f_k^* - E^*\right)+\left(E^*-0\right)+\left(\frac{1}{n}\sum_{k=1}^{n}(f_k-f_k^*)\right)\right|>\epsilon\right)$$

$$\leq P\left(\left|\left(\frac{1}{n}\sum_{k=1}^{n}f_k^* - E^*\right)+E^*\right|>\epsilon\right) + \sum_{k=1}^{n}P\left(f_k\neq f_k^*\right),$$

we shall estimate the two final expressions. From

$$E^* = \int\limits_{|x|<nC} x\,dF,$$

it follows that there is an integer N_1 such that $E^* < \epsilon/2$ for $n > N_1$. Since $D^2(f_k^*) < E((f_k^*)^2)$,

$$D^2(f_k^*) \leq nC \int\limits_{|x|<nC} |x|\,dF$$

and thus

$$D^2\left(\sum_{k=1}^{n}f_k^*\right) < \frac{n^2\delta\epsilon^2}{8}.$$

By Chebyshev's inequality with $\lambda = 1/\sqrt{(\delta/2)}$,

$$P\left(\left|\sum_{k=1}^{n}f_k^* - nE^*\right|>\frac{n\epsilon}{2}\right) < \frac{\delta}{2}$$

and so

$$P\left(\left|\frac{1}{n}\sum_{k=1}^{n}f_k^*\right|>\epsilon\right) < \frac{\delta}{2} \quad \text{for } n > N_1.$$

For the second expression, we have

$$P(f_k\neq f_k^*) = \int\limits_{|x|>nC} dF \leq \frac{1}{nC}\int\limits_{|x|>nC} |x|\,dF$$

and so

$$\sum_{k=1}^{n}P(f_k\neq f_k^*) \leq \frac{1}{C}\int\limits_{|x|>nC} |x|\,dF.$$

Since this last integral tends to 0 as $n \to \infty$, there is an integer $N > N_1$ such that

$$\int\limits_{|x|>nC} |x|\,dF < \frac{\delta C}{2} \quad \text{for } n > N.$$

Setting $N(\delta) = N$, we have established the required relation.

In contrast to the weak law of large numbers given by (2), a *strong law of large numbers* states that for n sufficiently large, the values of $(1/n)\sum f_k$ are confined to any given ϵ-neighborhood about E (except possibly on a set of measure zero); that is,

$$P\left(\limsup_{n\to\infty} \left| \frac{1}{n} \sum_{k=1}^{n} f_k - E \right| > \epsilon \right) = 0 \tag{4}$$

for any $\epsilon > 0$. Proofs of such strong laws use one or both of two basic approaches: "gap" methods and "high moment" methods. We shall not discuss the latter, since in our applications the variances may not exist. The essential idea of the former is to couple Chebyshev's inequality with the Borel-Cantelli lemma. In the development of (2), the probabilities were $\mathcal{O}(1/n)$ and hence the sum of these probabilities is divergent and we can not use the Borel-Cantelli lemma immediately. However, we can choose a subsequence of the events such that the sum of the corresponding probabilities is convergent and then, for these events at least, the Borel-Cantelli lemma will assure that (4) holds except for a set of measure zero. If we can show that the behavior of the sums $(1/n)\sum f_k$ for the n's in the gaps between the terms of our subsequence is not much different from that of the terms in the subsequence, we shall be finished. To make this outline more precise, let us prove:

Theorem 3. *Let f_1, f_2, ... be pairwise independent and identically distributed random variables with expectations $E = E(f_k)$ and variances $D^2 = D^2(f_k)$. Then*

$$P\left(\limsup_{n\to\infty} \left| \frac{1}{n} \sum_{k=1}^{n} f_k - E \right| > \epsilon \right) = 0$$

for any $\epsilon > 0$.

Proof. As before, Chebyshev's inequality gives that

$$P\left(\left| \frac{1}{n} \sum_{k=1}^{n} f_k - E \right| > \epsilon \right) < \frac{D^2}{n\epsilon^2},$$

so on the subsequence of squares $1, 2^2, 3^2, \ldots, n^2, \ldots$, we may apply the Borel-Cantelli lemma to find that

$$\left| \frac{1}{n^2} \sum_{k=1}^{n^2} f_k - E \right| \leq \epsilon$$

with probability one if n is sufficiently large, since the complementary event has probability zero. But what about the sums in the gaps; that is, for N's with $n^2 < N < (n+1)^2$? Clearly,

$$P\left(\max_{n^2 < N < (n+1)^2} \left| \frac{1}{n^2} \sum_{k=n^2+1}^{N} f_k - E \right| > \epsilon \right)$$

$$\leq \sum_{N=n^2+1}^{(n+1)^2-1} P\left(\left| \frac{1}{n^2} \sum_{k=n^2+1}^{N} f_k - E \right| > \epsilon \right),$$

and, by Chebyshev's inequality, this is

$$\leq \sum_{N=n^2+1}^{(n+1)^2-1} \frac{(N-n^2)D^2}{\epsilon^2 n^4} \leq \frac{4D^2}{\epsilon^2 n^2}. \tag{5}$$

Again applying the Borel-Cantelli lemma,

$$\max_{n^2 < N < (n+1)^2} \left| \frac{1}{n^2} \sum_{k=n^2+1}^{N} f_k - E \right| \leq \epsilon$$

with probability one if n is sufficiently large. But now for n sufficiently large, we see that for N with $n^2 < N < (n+1)^2$ we have

$$\left| \frac{1}{N} \sum_{k=1}^{N} f_k - E \right| \leq 2\epsilon$$

with probability one, and our result is established.

Since this proof depends directly on the convergence of $\sum 1/(\epsilon^2 n^2)$, a "strong law of large numbers with error term" immediately follows:

Corollary. *Let f_1, f_2, ... be pairwise independent and identically distributed random variables with expectations $E = E(f_k)$ and variances $D^2 = D^2(f_k)$. Then for all sufficiently large n we have that*

$$\left| \frac{1}{n} \sum_{k=1}^{n} f_k - E \right| \le \epsilon(n),$$

where $\epsilon(n)$ is any positive function decreasing to zero as $n \to \infty$ such that

$$\sum \frac{1}{\epsilon^2(n)\, n^2}$$

converges.

For example, $\epsilon(n)$ may be taken to be $\log(n)/\sqrt[4]{n}$ but, of course, there is no such "smallest" error function. It is clear that this "gap" method proof was successful because we could make the estimate (5) and apply the Borel-Cantelli lemma once again. In our later applications, we shall not have such a nice subsequence and so we will need a stronger estimate for the terms in the gaps, based on the following Kolmogorov-type inequality.

Theorem 4. *Let f_1, f_2, ... be totally independent and identically distributed random variables with expectations $E = E(f_k)$ and variances $D^2 = D^2(f_k)$. Let g_1, g_2, ... be $g_k = (f_1 + \cdots + f_k) - kE$. Then*

$$P\!\left(\max_{1 \le k \le n} g_k \ge x \right) \le \tfrac{4}{3}\, P\!\left(g_n \ge x - 2D\sqrt{n} \right)$$

for any real number x.

Proof. Let n be given, let S_k be the event that

$$g_1 < x,\ g_2 < x,\ \ldots,\ g_{k-1} < x \text{ and } g_k \ge x,$$

let T_k be the event that

$$g_n - g_k > -2D\sqrt{n}$$

and let S be the event that

$$g_n \ge x - 2D\sqrt{n}.$$

If both S_k and T_k occur, then S occurs as well. The events S_k and T_k are independent, since S_k depends on $g_1, \ldots, g_k$ while T_k depends on $g_{k+1}, \ldots, g_n$. Further, the events S_k are mutually exclusive, as are the events $S_k \cap T_k$. Since

$$\bigcup_{k=1}^{n} (S_k \cap T_k) \subseteq S,$$

the independence of S_k and T_k shows that

$$\sum_{k=1}^{n} P(S_k) \cdot P(T_k) = \sum_{k=1}^{n} P(S_k \cap T_k) \leq P(S).$$

But

$$1 - P(T_k) \leq P(\,|g_n - g_k| \geq 2D\sqrt{n}\,)$$

and so, by Chebyshev's inequality,

$$1 - P(T_k) \leq \frac{n-k}{4n} \leq \frac{1}{4}.$$

Hence $P(T_k) \geq 3/4$,

$$\frac{3}{4} \sum_{k=1}^{n} P(S_k) \leq P(S)$$

and

$$P(\max_{1 \leq k \leq n} g_k \geq x) \leq \frac{4}{3} P(S),$$

as required.

§5. The Gauss-Kuzmin theorem.

In his letter to Laplace of 30 January 1812, Gauss described a "curious problem" that had occupied him for twelve years and that he was unable to resolve to his satisfaction. Let $0 \leq x \leq 1$ and let $m_n(x)$ be the probability that the number $t \in [0,1)$ has $1/\zeta_{n+1}(t) < x$. It is clear that $m_0(x) = x$ and that m_{n+1} depends on m_n since $\zeta_{n+1} = a_{n+1} + 1/\zeta_{n+2}$. Although he does not explicitly state it, Gauss must have known the recursion formula

$$m_{n+1}(x) = \sum_{k=1}^{\infty} \left(m_n(\tfrac{1}{k}) - m_n(\tfrac{1}{k+x}) \right),$$

since he wrote that he could prove by a very simple argument that the limit as n $\to \infty$ of $m_n(x)$ is $\log(1+x)/\log(2)$. If this limit does exist, it must satisfy the functional equation

$$m(x) \;=\; \sum_{k=1}^{\infty}\Big(m(\tfrac{1}{k}) \;-\; m(\tfrac{1}{k+x})\Big)$$

and it also must be 0 at $x = 0$ and 1 at $x = 1$. The function $\log(1+x)/\log(2)$ has these properties and it is possible that these facts led Gauss to his proof. Gauss stated that he was unable to describe the difference $m_n(x) - \log(1+x)/\log(2)$ for large n.

More than a century later, Kuzmin [1928] established that $m_n(x) = \log(1+x)/\log(2) + \mathcal{O}(q^{\sqrt{n}})$ where the constant q satisfies $0 < q < 1$. At nearly the same time, Lévy [1929] was able to obtain $\mathcal{O}(q^n)$ with $q = 0.7$ by a completely different method. We will give a simple proof that

$$m_n(x) \;=\; \frac{\log(1+x)}{\log(2)} + \mathcal{O}(q^n),$$

where $0 < q < 1$ satisfies $q < 0.76$. In fact, we will show more:

Theorem 1. *Let $f_0(x)$ be any twice differentiable function defined on $[0,1]$ with $f_0(0) = 0$ and $f_0(1) = 1$. Let the sequence of functions $f_1(x)$, $f_2(x)$, ... be defined by the recursion formula*

$$f_{n+1}(x) \;=\; \sum_{k=1}^{\infty}\Big(f_n(\tfrac{1}{k}) \;-\; f_n(\tfrac{1}{k+x})\Big).$$

Then

$$f_n(x) \;=\; \frac{\log(1+x)}{\log(2)} \;+\; \mathcal{O}(q^n),$$

where $0 < q < 1$ satisfies $q < 0.76$.

It is clear that for $f_0(x) = x = m_0(x)$, this theorem will establish Gauss' claim and provide an answer to his problem.

Proof. Instead of studying $f_n(x)$ directly, we look at the derivative

$$f'_{n+1}(x) = \sum_{k=1}^{\infty} \frac{1}{(k+x)^2}\, f'_n\Big(\frac{1}{k+x}\Big). \tag{1}$$

Let us introduce another sequence of functions g_0, g_1, ... defined by

$$f'_n(x) = \frac{g_n(x)}{1+x}.$$

Then the recursion formula (1) is transformed into

$$\frac{g_{n+1}(x)}{1+x} = \sum_{k=1}^{\infty} \frac{1}{(k+x)^2}\, \frac{g_n\big(\frac{1}{k+x}\big)}{1+\frac{1}{k+x}} = \sum_{k=1}^{\infty} g_n\Big(\frac{1}{k+x}\Big) \frac{1}{(k+x)(k+x+1)}\,,$$

or

$$g_{n+1}(x) = \sum_{k=1}^{\infty} g_n\Big(\frac{1}{k+x}\Big) \frac{1+x}{(k+x)(k+x+1)}.$$

If we can show that $g_n(x) = 1/\log(2) + \mathcal{O}(q^n)$, then an integration will establish the theorem for $f_n(x)$, because integrating $1/(1+x)$ will give the $\log(1+x)$ term together with a bounded expression on a bounded interval times the $\mathcal{O}(q^n)$ error term, which will remain $\mathcal{O}(q^n)$. To demonstrate that $g_n(x)$ has this desired form, it suffices to establish that $g'_n(x) = \mathcal{O}(q^n)$, as the $1/\log(2)$ constant in $g_n(x)$ will follow from the normalization requirement that $f_0(0) = 0$ and $f_0(1) = 1$.

We find that

$$g'_{n+1}(x) = -\sum_{k=1}^{\infty} g'_n\Big(\frac{1}{k+x}\Big) \frac{1+x}{(k+x)^3(k+x+1)}$$

$$+ \sum_{k=1}^{\infty} g_n\Big(\frac{1}{k+x}\Big) \frac{k(k-1)-(1+x)^2}{(k+x)^2(k+x+1)^2}.$$

Since

$$\sum_{k=1}^{\infty} \frac{1+x}{(k+x)(k+x+1)} = (1+x)\sum_{k=1}^{\infty} \Big(\frac{1}{k+x} - \frac{1}{k+x+1}\Big) \equiv 1$$

and the term by term differentiation is justified, we have immediately that

$$\Big(\sum_{k=1}^{\infty} \frac{1+x}{(k+x)(k+x+1)}\Big)' = -\sum_{k=1}^{\infty} \frac{k(k-1)-(1+x)^2}{(k+x)^2(k+x+1)^2} \equiv 0,$$

and so

$$\sum_{k=1}^{\infty} g_n\left(\frac{1}{1+x}\right) \frac{k(k-1)-(1+x)^2}{(k+x)^2(k+x+1)^2} = 0.$$

Now we may rewrite our formula for $g'_{n+1}(x)$ as

$$g'_{n+1}(x) = -\sum_{k=1}^{\infty} g'_n\left(\frac{1}{k+x}\right) \frac{1+x}{(k+x)^3(k+x+1)}$$

$$- \sum_{k=1}^{\infty} \left(g_n\left(\frac{1}{1+x}\right) - g_n\left(\frac{1}{k+x}\right)\right) \frac{k(k-1)-(1+x)^2}{(k+x)^2(k+x+1)^2}.$$

By applying the mean value theorem of calculus to the difference

$$g_n\left(\frac{1}{1+x}\right) - g_n\left(\frac{1}{k+x}\right),$$

we obtain

$$g'_{n+1}(x) = -\sum_{k=1}^{\infty} g'_n\left(\frac{1}{k+x}\right) \frac{1+x}{(k+x)^3(k+x+1)}$$

$$- \sum_{k=2}^{\infty} g'_n\left(\frac{1}{\theta_k+x}\right) \frac{k-1}{1+x} \frac{k(k-1)-(1+x)^2}{(k+x)^3(k+x+1)^2},$$

where $1 < \theta_k < k$.

Let M_n be the maximum of $|g'_n(x)|$ on $x \in [0,1]$. Then our expression for $g'_{n+1}(x)$ shows that M_{n+1} is not greater than

$$M_n \cdot \max_{x \in [0,1]} \left(\sum_{k=1}^{\infty} \frac{1+x}{(k+x)^3(k+x+1)} + \sum_{k=2}^{\infty} \left| \frac{(k-1)\left(k(k-1)-(1+x)^2\right)}{(1+x)(k+x)^3(k+x+1)^2} \right| \right).$$

We now must calculate the maximum value of the sums in this expression. We can drop the absolute value signs from these sums because all the terms are positive for $x \in [0,1]$ except for the first term of the second sum, which takes its maximum value when $x = 0$. After combining the resulting terms, we have to find

$$\max_{x \in [0,1]} \sum_{k=1}^{\infty} \frac{(1+x)^2(2x) + k(k-1)^2}{(1+x)(k+x)^3(k+x+1)^2}.$$

By logarithmic differentiation, the derivative of one of the terms of this sum is

$$\frac{(1+x)(3+5x)-\left(\dfrac{1}{1+x}+\dfrac{3}{k+x}+\dfrac{2}{k+1+x}\right)\left((1+x)^2(2+x)+k(k-1)^2\right)}{(1+x)(k+x)^3(k+1+x)^2}$$

and this expression is negative on $[0,1]$ for $k = 1, 2, \dots$. So the maximum of our expression must occur at $x = 0$ and we have that

$$M_{n+1} \le M_n \sum_{k=1}^{\infty} \frac{2+k(k-1)^2}{k^3(k+1)^2}.$$

Partial fraction decomposition gives

$$\sum_{k=1}^{\infty} \frac{2+k(k-1)^2}{k^3(k+1)^2} = \sum_{k=1}^{\infty}\left(\frac{2}{k^3}-\frac{1}{k^2}\right)+2\sum_{k=1}^{\infty}\left(\left(\frac{1}{k}-\frac{1}{k+1}\right)-\left(\frac{1}{k^2}-\frac{1}{(k+1)^2}\right)\right)$$

$$= 2\zeta(3)-\zeta(2) < 0.76,$$

where $\zeta(n)$ denotes the Riemann zeta function. Thus $M_{n+1} < M_n \cdot q$, where $q < 0.76$, and $g'_{n+1}(x) = \mathcal{O}(q^{n+1})$, which proves the theorem.

We may use our formula for $m_n(x)$ to make a precise calculation of $P(a_{n+1} = k)$, which we had estimated in (3.2). Since $a_{n+1} = k$ is equivalent to $k \le \zeta_{n+1} < k+1$, $P(a_{n+1} = k)$ is the same as $m_n(1/k) - m_n(1/(k+1))$. Thus, by the proof of Theorem 1,

$$P(a_{n+1}=k) = \int_{1/(k+1)}^{1/k} \left(\frac{1}{\log(2)}\frac{1}{1+x}+\mathcal{O}(q^n)\right)dx$$

$$= \frac{1}{\log(2)}\log\left(1+\frac{1}{k(k+2)}\right)+\mathcal{O}\left(\frac{q^n}{k(k+1)}\right). \tag{2}$$

From the power series expansion of $\log(1+x)$, we see that this new expression satisfies the weaker estimate in (3.2).

§6. The distribution of ξ_n.

Let $0 \le x \le 1$ and let $m_n^*(x)$ be the probability that $t \in [0,1)$ has $\xi_n < x$. Lévy's solution of Gauss's problem also established that

$$m_n^*(x) = \frac{\log(1+x)}{\log(2)} + \mathcal{O}(q^n),$$

where the constant q satisfies $0 < q < 1$. Essentially, Lévy [1929] observed that $m_n(x)$ and $m_n^*(x)$ coincide at "many" rational values and then showed that the difference between them at the intermediate values is small. In this section, we show how this formula for $m_n^*(x)$ follows without reference to Theorem 5.1.

Theorem 1. $m_n^*(x) = \log(1+x)/\log(2) + \mathcal{O}(q^n)$ *where* $0 < q < 1$ *satisfies* $q < 47/60$.

Proof. We shall find a recursion relation between m_{n+1}^* and m_n^*. We begin with a connection between m_n and m_n^*. For a given interval I_n of order n, those numbers with $1/\zeta_{n+1} < x$ form a sub-interval of length

$$\left| \frac{A_n(\frac{1}{x}) + A_{n-1}}{B_n(\frac{1}{x}) + B_{n-1}} - \frac{A_n}{B_n} \right| = \frac{x}{B_n(B_n + B_{n-1}x)}$$

and so, by (1.1), the fraction of I_n with $1/\zeta_{n+1} < x$ is

$$\frac{B_n(B_n + B_{n-1})x}{B_n(B_n + B_{n-1}x)} = \frac{(1 + \xi_n)x}{1 + \xi_n x}. \tag{1}$$

Thus

$$m_n(x) = \sum_{k_1, \ldots, k_n} |I_n| \cdot \frac{(1 + \xi_n)x}{1 + \xi_n x},$$

where the summation is over all intervals of order n as in (1.2). Since each $I_n = I_n(k_1, \ldots, k_n)$ corresponds to a particular value of $\xi_n = [0; k_n, \ldots, k_1]$ and $m_n^*(x)$ is a step function, we have that

$$m_n(x) = \int_0^1 \frac{(1+t)x}{1+tx} \, dm_n^*(t).$$

We now turn to $m_{n+1}^*(x) = P(\xi_{n+1} < x)$. Since $\xi_{n+1} = 1/(a_{n+1} + \xi_n)$, we have that

$$m_{n+1}^*(x) = P(a_{n+1} \geq [\tfrac{1}{x}] + 1) + P(a_{n+1} = [\tfrac{1}{x}] \text{ and } \xi_n > \{\tfrac{1}{x}\}).$$

The integer bound in the first term allows us to replace $a_{n+1} \geq [1/x] + 1$ by the condition $\zeta_{n+1} > [1/x] + 1$ and so

$$m_{n+1}^*(x) = m_n\left(\frac{1}{[\frac{1}{x}]+1}\right) + P(a_{n+1} = [\tfrac{1}{x}] \text{ and } \xi_n > \{\tfrac{1}{x}\})$$

$$= \int_0^1 \frac{1+t}{[\frac{1}{x}]+1+t}\, dm_n^*(t) + P(a_{n+1} = [\tfrac{1}{x}] \text{ and } \xi_n > \{\tfrac{1}{x}\}).$$

For a given interval I_n of order n, those numbers with $a_{n+1} = [1/x]$ form a subinterval of length

$$\left| \frac{A_n([\frac{1}{x}]) + A_{n-1}}{B_n([\frac{1}{x}]) + B_{n-1}} - \frac{A_n([\frac{1}{x}]+1) + A_{n-1}}{B_n([\frac{1}{x}]+1) + B_{n-1}} \right|$$

$$= \frac{1}{\left(B_n([\frac{1}{x}]) + B_{n-1}\right)\left(B_n([\frac{1}{x}]+1) + B_{n-1}\right)}$$

and, just as for (1), the fraction of I_n with $a_{n+1} = [1/x]$ is

$$\frac{1+\xi_n}{\left([\frac{1}{x}]+\xi_n\right)\left([\frac{1}{x}]+1+\xi_n\right)} = \frac{1+\xi_n}{[\frac{1}{x}]+\xi_n} - \frac{1+\xi_n}{[\frac{1}{x}]+1+\xi_n}.$$

Summing over all intervals I_n of order n such that $\xi_n > \{1/x\}$, we obtain

$$P(a_{n+1} = [\tfrac{1}{x}] \text{ and } \xi_n > \{\tfrac{1}{x}\}) = \int_{\{1/x\}}^1 \left(\frac{1+t}{[\frac{1}{x}]+t} - \frac{1+t}{[\frac{1}{x}]+1+t} \right) dm_n^*(t)$$

and thus we find the recursion formula

$$m_{n+1}^*(x) = \int_0^{\{1/x\}} \frac{1+t}{[\frac{1}{x}]+1+t}\, dm_n^*(t) + \int_{\{1/x\}}^1 \frac{1+t}{[\frac{1}{x}]+t}\, dm_n^*(t).$$

Because the function $\log(1+x)$ satisfies the function equation

$$m^*(x) = \int_0^{\{1/x\}} \frac{1+t}{[\frac{1}{x}]+1+t}\, dm^*(t) + \int_{\{1/x\}}^1 \frac{1+t}{[\frac{1}{x}]+t}\, dm^*(t)$$

corresponding to our recursion relation for $m_{n+1}^*(x)$, we now have that

$$E_{n+1}^*(x) = \int_0^{\{1/x\}} \frac{1+t}{[\frac{1}{x}]+1+t}\, dE_n^*(t) + \int_{\{1/x\}}^1 \frac{1+t}{[\frac{1}{x}]+t}\, dE_n^*(t),$$

where the error functions $E_n^*(x)$ are

$$E_n^*(x) \;=\; m_n^*(x) \;-\; \frac{\log(1+x)}{\log(2)}.$$

Since $E_n^*(0) = 0 = E_n^*(1)$, integration by parts gives that

$$\int_0^{\{1/x\}} \frac{1+t}{[\frac{1}{x}]+1+t}\, dE_n^*(t) \;=\; \frac{1+\{\frac{1}{x}\}}{\frac{1}{x}+1}\, E_n^*(\{\tfrac{1}{x}\}) \;-\; \int_0^{\{1/x\}} E_n^*(t)\left(\frac{1+t}{[\frac{1}{x}]+1+t}\right)' dt$$

and

$$\int_{\{1/x\}}^1 \frac{1+t}{[\frac{1}{x}]+t}\, dE_n^*(t) \;=\; -\frac{1+\{\frac{1}{x}\}}{\frac{1}{x}}\, E_n^*(\{\tfrac{1}{x}\}) \;-\; \int_{\{1/x\}}^1 E_n^*(t)\left(\frac{1+t}{[\frac{1}{x}]+t}\right)' dt.$$

Let $y = (1 + \{1/x\})(x - x/(x+1))$ for $0 < x < 1$. If $0 < x < 1/2$, then $1 + \{1/x\} < 2$ and $x - x/(x+1) < 1/6$ so that $y < 1/3$. If $(k-1)/k < x < k/(k+1)$, where $k > 1$ is an integer, then $1 + 1/k < 1/x < 1 + 1/(k-1)$ and so

$$y \;<\; (1 + \frac{1}{k-1})(\frac{k^2}{(k+1)(2k+1)}) \;=\; \frac{k^3}{(k^2-1)(2k+1)}.$$

Thus $y < 1/2$ for $k \geq 3$, $y < 8/15$ for $k = 2$, and hence

$$(1 + \{\tfrac{1}{x}\})(x - \frac{x}{x+1}) \;<\; \tfrac{8}{15} \quad \text{for} \quad 0 < x < 1.$$

Now let $y = (1+t)/(k+t)$ where k is a positive integer. Then $y' = (k-1)/(k+t)^2$ is non-increasing on $[0,1]$ and we find that

$$\max_t \left(\frac{1+t}{k+t}\right)' \;\leq\; \tfrac{1}{4}.$$

Applying these estimates to the results of the integrations by parts, we have for any x in $[0,1]$ that

$$|E_{n+1}^*(x)| \;\leq\; \left(\max_t |E_n^*(t)|\right) \cdot \left(\tfrac{8}{15} + \tfrac{1}{4}(\{\tfrac{1}{x}\} - 0) + \tfrac{1}{4}(1 - \{\tfrac{1}{x}\})\right),$$

and the proof is completed with $q = 47/60$.

§7. Partial quotients are weakly dependent.

While the improved estimate of $P(a_{n+1} = k)$ in (5.2) follows from the Gauss-Kuzmin theorem, the more general nature of Theorem 5.1 allows us to show that the partial quotients of almost all $t \in [0, 1)$ are weakly dependent.

Theorem 1. *Let n, k, r and s be positive integers. Then*

$$P(a_n = r \text{ and } a_{n+k} = s) = P(a_n = r) \cdot P(a_{n+k} = s) \cdot \left(1 + \mathcal{O}(q^k)\right), \qquad (1)$$

where $0 < q < 1$.

Proof. For a given interval I_n of order n, let $0 \le x \le 1$ and let $M_k(x)$ be the probability that $t \in I_n$ has $1/\zeta_{n+k+1} < x$. As with the Gauss measure $m_n(x)$, we see immediately that

$$M_{k+1}(x) = \sum_{j=1}^{\infty} \left(M_k\left(\frac{1}{j}\right) - M_k\left(\frac{1}{j+x}\right) \right).$$

In order to apply Theorem 5.1, we normalize $M_k(x)$ by setting $f_k(x) = M_k(x)/|I_n|$ so that $f_k(0) = 0$ and $f_k(1) = 1$. Since $f_0(x)$ is the fraction of I_n with $1/\zeta_{n+1} < x$, we know from (6.1) that

$$f_0(x) = \frac{(1+\xi_n)x}{1+\xi_n x}$$

and, as this rational function is twice differentiable on $[0, 1]$, the sequence of functions $f_0(x)$, $f_1(x)$, ... satisfies the requirements of Theorem 5.1. It follows from (5.1) that

$$f_k'(x) = \frac{1}{\log(2)} \frac{1}{1+x} + \mathcal{O}(q^k).$$

Thus

$$P(t \in I_n \text{ has } a_{n+k} = s) = |I_n| \int_{1/(s+1)}^{1/s} \left(\frac{1}{\log(2)} \frac{1}{1+x} + \mathcal{O}(q^k) \right) dx$$

$$= |I_n| \left(\frac{1}{\log(2)} \log\left(1 + \frac{1}{s(s+2)}\right) + \mathcal{O}\left(\frac{q^k}{s(s+1)}\right) \right).$$

Summing over all intervals I_n of order n with $a_n = r$,

$$P(a_n = r \text{ and } a_{n+k} = s) = P(a_n = r) \cdot \left(\frac{1}{\log(2)} \log\left(1 + \frac{1}{s(s+2)}\right) + \mathcal{O}\left(\frac{q^k}{s(s+1)}\right) \right)$$

and (1) follows from (5.2) and the estimate that

$$P(a_{n+k} = s) = \mathcal{O}\left(\frac{1}{s(s+1)}\right),$$

which follows from (3.2).

§8. Khintchine's theorem.

Given the weak dependence of the partial quotients, we may prove a strong law of large numbers due to Khintchine [1935] by applying the method of Theorem 4.3. As we shall need both expectations and variances to exist, we shall restrict our attention to functions $f(k)$ that increase slowly enough that the first and second moments (that is, $\sum f(k) \cdot P(a_n = k)$ and $\sum f^2(k) \cdot P(a_n = k)$ in the same manner as (3.1)) exist. Since the error terms in (5.2) and (7.1) are constants times powers of q, sums of such expressions will be geometric series and therefore will be bounded above by absolute constants. Thus the only essential change in the proof of Theorem 4.3 will be the values of the constants, and these values are irrelevant to the final results.

Theorem 1 ("Khintchine's Theorem"). *Let $f(r) = \mathcal{O}(r^p)$, where $0 \leq p < 1/2$, be a function defined on the positive integers. Then for all sufficiently large n we have that*

$$\left| \frac{1}{n} \sum_{k=1}^{n} f(a_k) - \frac{1}{\log(2)} \sum_{r=1}^{\infty} f(r) \log\left(1 + \frac{1}{r(r+2)}\right) \right| \leq \epsilon(n)$$

for almost all $t \in [0,1)$, where the error function $\epsilon(n)$ is any positive function decreasing to zero as $n \to \infty$ such that $\sum 1/(\epsilon^2(n) \cdot n^2)$ converges.

Proof. Let $f_k = f(a_k)$ and consider $\sum f_k$ for a particular value of n. By (5.2), the expectation E_k of f_k is

$$\sum_{r=1}^{\infty} f(r) \left(\frac{1}{\log(2)} \log\left(1 + \frac{1}{r(r+2)}\right) + \mathcal{O}\left(\frac{q^k}{r(r+1)}\right) \right)$$

and thus

$$E_k = \frac{1}{\log(2)} \sum_{r=1}^{\infty} f(r) \log\left(1 + \frac{1}{r(r+2)}\right) + \mathcal{O}(q^k),$$

since $\sum f(r)/(r(r+1))$ is convergent. Hence

$$\sum_{k=1}^{n} E_k = \frac{n}{\log(2)} \sum_{r=1}^{\infty} f(r) \log\left(1 + \frac{1}{r(r+2)}\right) + \mathcal{O}(1),$$

since the error terms form a geometric series. For the variances, we look first at the second moments.

$$\int_0^1 f_k^2(t) \, dt = \frac{1}{\log(2)} \sum_{r=1}^{\infty} f^2(r) \left(\frac{1}{\log(2)} \log\left(1 + \frac{1}{r(r+2)}\right) + \mathcal{O}\left(\frac{q^k}{r(r+1)}\right) \right)$$

exists, since $f^2(r) = \mathcal{O}(r^{2p})$ and $2p < 1$. Setting

$$E = \sum_{k=1}^{n} E_k,$$

we find

$$D^2\left(\sum_{k=1}^{n} f_k\right) = \int_0^1 \left(\sum_{k=1}^{n} f_k\right)^2 dt - E^2$$

$$= \sum_{k=1}^{n} \int_0^1 f_k^2 \, dt + 2\sum_{k=1}^{n-1} \sum_{j=1}^{n-k} \int_0^1 f_k \cdot f_{k+j} \, dt - \sum_{k=1}^{n} E_k^2 - 2\sum_{k=1}^{n-1} \sum_{j=1}^{n-k} E_k \cdot E_{k+j}$$

$$= \sum_{k=1}^{n} \int_0^1 f_k^2 \, dt + 2\sum_{k=1}^{n-1} \sum_{j=1}^{n-k} \left(\int_0^1 f_k f_{k+j} \, dt - E_k E_{k+j} \right) - \sum_{k=1}^{n} E_k^2.$$

Since the partial quotients are weakly dependent, we can estimate the terms of the double sum by using (7.1):

$$\int_0^1 f_k f_{k+j} \, dt - E_k E_{k+j}$$

$$= \sum_{r=1}^{\infty} \sum_{s=1}^{\infty} f(r) f(s) \left(P(a_k = r \text{ and } a_{k+j} = s) - P(a_k = r) \cdot P(a_{k+j} = s) \right)$$

$$= \sum_{r=1}^{\infty} \sum_{s=1}^{\infty} f(r)f(s) \left(P(a_k = r) \cdot P(a_{k+j} = s) \cdot \mathcal{O}(q^j) \right) = E_k E_{k+j} \cdot \mathcal{O}(q^j).$$

Thus

$$D^2\left(\sum_{k=1}^{n} f_k \right) = \sum_{k=1}^{n} \int_0^1 f_k^2 \, dt + \mathcal{O}\left(\sum_{k=1}^{n-1} E_k \cdot \sum_{j=1}^{n-k} E_{k+j} \cdot q^j \right) - \sum_{k=1}^{n} E_k^2$$

$$= \sum_{k=1}^{n} \int_0^1 f_k^2 \, dt + \mathcal{O}\left(\sum_{k=1}^{n} E_k^2 \right) = \mathcal{O}(n),$$

since, by Schwarz's inequality, $D^2(f_k) = \mathcal{O}(E(f_k^2))$. We now have a constant $K > 0$ that is independent of n and such that $D < K\sqrt{n}$. Again by Chebyshev's inequality,

$$P\left(\left| \sum_{k=1}^{n} f_k - nE \right| > \epsilon n \right) < \frac{K^2}{\epsilon^2 n}$$

for any $\epsilon > 0$. The remainder of the proof proceeds as in Theorem 4.3 and its Corollary.

Let us investigate two particular choices of the function $f(r)$ in Khintchine's theorem. First, let $f(r)$ be the indicator function of the positive integer N; that is, $f(r) = 1$ if $r = N$ and $f(r) = 0$ otherwise. Then

$$\frac{1}{n} \sum_{k=1}^{n} f(a_k) = \frac{1}{n} \sum_{\substack{1 \le k \le N \\ a_k = N}} 1$$

is the *relative frequency* of the digit N in the first n partial quotients of the continued fraction. Theorem 1 gives that the limit as $n \to \infty$ of this relative frequency is

$$\frac{1}{\log(2)} \log\left(1 + \frac{1}{N(N+2)} \right).$$

Thus the number N occurs with the same relative frequency in the continued fractions of almost all numbers.

If we let $f(r) = \log(r)$, then

$$\frac{1}{n} \sum_{k=1}^{n} f(a_k) = \frac{1}{n} \sum_{k=1}^{n} \log(a_k) = \frac{1}{n} \log(a_1 \cdot \cdots \cdot a_n)$$

is the logarithm of the *geometric mean* of the first n partial quotients. By Khintchine's theorem, for all sufficiently large n we have that

$$\left| \log\left(\sqrt[n]{a_1 \cdot \ \cdots \ \cdot a_n} \right) - \frac{1}{\log(2)} \sum_{r=1}^{\infty} \log(r) \log\left(1 + \frac{1}{r(r+2)}\right) \right| \leq \epsilon(n)$$

and so

$$\sqrt[n]{a_1 \cdot \ \cdots \ \cdot a_n} \ \rightarrow \ \prod_{r=1}^{\infty} \left(1 + \frac{1}{r(r+2)} \right)^{\frac{\log(r)}{\log(2)}}$$

as $n \rightarrow \infty$ for almost all numbers. This absolute constant is known as *Khintchine's constant*. Since

$$B_{n+1} = a_{n+1} B_n + B_{n-1} < 2 a_{n+1} B_n = 2 a_{n+1}(a_n B_{n-1} + B_{n-2})$$

$$< 2^2 \, a_{n+1} a_n \, B_{n-1} < \cdots < 2^n \, a_{n+1} \cdot \ \cdots \ \cdot a_1 \, B_0,$$

we have immediately from the above that

$$\sqrt[n]{B_n} \ = \ \mathcal{O}(1) \tag{1}$$

for almost all numbers.

§9. The Khintchine-Lévy theorem for $\sqrt[n]{B_n}$.

Given Khintchine's result that for almost all numbers the n-th root of the n-th denominator is bounded by an absolute constant, it is natural to ask if there is an expected value for $\sqrt[n]{B_n}$ and then, if there is such a value, to ask for the corresponding law of large numbers. Khintchine [1936] established that the limit of $\sqrt[n]{B_n}$ exists independently of t almost everywhere and, at about the same time, Lévy [1936] also established the existence of the limit and determined its value.

Theorem 1. *Let* $\gamma = \dfrac{\pi^2}{12 \log(2)}$. *Then for all sufficiently large n we have that*

$$\left| \frac{1}{n} \log(B_n) - \gamma \right| \leq \epsilon(n) \tag{1}$$

for almost all $t \in [0,1)$, *where the error function* $\epsilon(n)$ *is any positive function decreasing to zero as* $n \to \infty$ *such that* $\sum 1/(\epsilon^2(n) \cdot n^2)$ *converges.*

Thus we find that $\sqrt[n]{B_n} \to \exp(\gamma)$ as $n \to \infty$ for almost all numbers. The constant γ is known as the *Khintchine-Lévy constant.*

Proof. We shall establish this result by showing a strong law of large numbers for the random variables $\log(\zeta_k)$, using again the method of Theorem 4.3 and its Corollary. From (I.4.3) and (I.4.4),

$$\zeta_{n+2} = \frac{B_{n+1}\zeta_{n+2} + B_n}{B_n\zeta_{n+1} + B_{n-1}}$$

so that

$$\zeta_{n+2} \cdot \cdots \cdot \zeta_2 = \frac{B_{n+1}\zeta_{n+2} + B_n}{B_0\zeta_1 + B_{-1}}$$

and then, since $B_0 = 1$ and $B_{-1} = 0$, we have

$$\zeta_{n+2} \cdot \cdots \cdot \zeta_1 = B_{n+1}\zeta_{n+2} + B_n$$

or

$$\zeta_1 \cdot \cdots \cdot \zeta_{n+1} = B_{n+1}\left(1 + \frac{B_n}{\zeta_{n+2}B_{n+1}}\right).$$

Taking logarithms,

$$\frac{1}{n+1}\sum_{k=1}^{n+1}\log(\zeta_k) = \frac{1}{n+1}\log(B_{n+1}) + \frac{1}{n+1}\log\left(1 + \frac{\zeta_{n+1}}{\zeta_{n+2}}\right)$$

and we see that the behavior of $\log(B_{n+1})/(n+1)$ is asymptotic to the right-hand sum as n increases.

We compute the expectation of $\log(\zeta_n)$ as follows:

$$\int_0^1 \log(\zeta_n)\, dt = \frac{1}{\log(2)}\int_0^1 \log\left(\tfrac{1}{t}\right)\left(1 + \mathcal{O}(q^n)\right)\frac{dt}{1+t},$$

by the proof of Theorem 5.1, and this is the same as

$$\frac{1}{\log(2)}\int_1^\infty \log(y)\left(1 + \mathcal{O}(q^n)\right)\frac{dy}{y^2\left(1 + \frac{1}{y}\right)}$$

by setting $y = 1/t$, so that $dt = -dy/y^2$, and then reversing the limits of integration. Expanding $1/(1 + 1/y)$ as a power series in $1/y$, we have

$$\frac{1}{\log(2)} \sum_{k=0}^{\infty} (-1)^k \int_1^{\infty} \frac{\log(y)}{y^{k+2}} \, dy + \mathcal{O}(q^n).$$

Integrating by parts, we find

$$\int_1^{\infty} \log(y) \frac{dy}{y^{k+2}} = \frac{-1}{k+1} \frac{\log(y)}{y^{k+1}} \Big|_1^{\infty} + \frac{1}{k+1} \int_1^{\infty} \frac{dy}{y^{k+1}} = 0 + \frac{1}{(k+1)^2}.$$

Noticing that

$$\sum_{k=1}^{\infty} \frac{(-1)^k}{k^2} = \sum_{k=1}^{\infty} \frac{1}{k^2} - 2 \sum_{k=1}^{\infty} \frac{1}{(2k)^2} = \zeta(2) - \frac{\zeta(2)}{2},$$

where $\zeta(2) = \pi^2/6$ is the value of the Riemann zeta function at 2, we obtain

$$\int_0^1 \log(\zeta_n) \, dt = \gamma + \mathcal{O}(q^n).$$

Thus

$$E = \int_0^1 \left(\sum_{k=1}^{n+1} \log(\zeta_k) \right) dt = (n+1)\gamma + \mathcal{O}(1).$$

For the variances, we have

$$D^2 \left(\sum_{k=1}^{n+1} \log(\zeta_k) \right) = \int_0^1 \left(\sum_{k=1}^{n+1} \log(\zeta_k) \right)^2 dt - E^2$$

as in the proof of Theorem 8.1. Therefore we must evaluate

$$\sum_{k=1}^{n+1} \int_0^1 \log^2(\zeta_k) \, dt + 2 \sum_{k=1}^{n} \sum_{j=1}^{n+1-k} \int_0^1 \log(\zeta_k) \log(\zeta_{k+j}) \, dt - E^2.$$

For the terms of the double sum, we have as above that

$$\int_0^1 \log(\zeta_k) \log(\zeta_{k+j}) \, dt = \int_0^1 \log(x) \log(y) \, dF(x,y),$$

where $F(x,y)$ is the distribution function $P(1/\zeta_k < x \text{ and } 1/\zeta_{k+j} < y)$. If we had that

$$dF(x,y) = dm_{k-1}(x) \cdot dm_{k+j-1}(y) \cdot (1 + \mathcal{O}(q^j)),$$

then we could proceed, as in the proof of Khintchine's theorem, to show that $D^2 = \mathcal{O}(1/(n+1))$ and then complete the proof with the usual gap method. Since ζ_{k+j} depends on ζ_k, this is not the case. However, if for a large positive integer N we set

$$\zeta_k^* = [a_k; a_{k+1}, \ldots, a_{k+N}] \quad \text{for } k = 1, \ldots, n+1,$$

then ζ_k^* and ζ_{k+j}^* will have the desired relationship. Given any $\epsilon > 0$, $|\zeta_k - \zeta_k^*| < \epsilon$ for N sufficiently large, and the proof is finished.

NOTES

§1 through §4. Much of our presentation of basic probability theory is adapted from Rényi [1970]; see also Révész [1968]. Refinements in the gap method and error estimate of Theorem 4.3 and its Corollary are well-known (see Chapter VII of Rényi [1970]). Our "totally" independent random variables are also called "mutually" independent by some authors.

§5. See Szüsz [1961]. The letter from Gauss to Laplace is also reproduced in Uspensky [1937], Appendix III. While sufficient for our purposes, Theorem 1 can be improved in two directions. First, it is possible to relax the differentiability requirements on the initial function. Szüsz [1968] showed that one can take $f_0(x)$ to be continuously differentiable and the "speed" of the convergence depends only on the modulus of continuity of $f_0'(x)$. Second, the value for q can be significantly reduced. By a more careful analysis, Szüsz [1961] obtained $q \leq 0.485$. Wirsing [1974] refined Szüsz's approach still further and discussed the minimal value for q. He studied the recursion formula for f' as an operator on the space of continuous functions on $[0,1]$, found the first eigenvalue, and gave an approximation to the eigenfunctions. The complete spectrum of this operator was determined by Babenko [1978] and some computational results may be found in Babenko and Jur'ev [1978]. See also Waterman [1971].

§6. See also Lévy [1937], Chapter IX, and §76 in particular.

§7. The weak dependence of the partial quotients means that they form a "mixing sequence of random variables." Philipp [1971], Chapter 2, finds the corresponding "central limit theorem" and "law of iterated logarithm" with error estimates (see also Philipp and Stout [1975]). Billingsley [1968] gives the central limit theorem (but no error estimate) and also derives an "arc sine law for best approximations."

§8 and §9. The limit relations (that is, without error estimates) of these sections can also be established by application of the "ergodic theorem" to the transformation T (see §3 of Chapter I) and its invariant measure $m(x)$; see Ryll-Nardzewski [1951], Kac [1959], Billingsley [1965], Chapter 1, §4, and also Series [1982].

The use of ζ_k^* as an approximation to ζ_k in §9 originates with Khintchine [1936].

Chapter VI

APPLICATIONS TO
METRICAL DIOPHANTINE APPROXIMATION

In Chapter II, we examined the solutions of the diophantine approximation $b \parallel bt \parallel < K$ for a constant $K > 0$ and, in Chapter IV, we saw that for numbers with bounded partial quotients, there were no solutions for sufficiently small values of K. As we now know from Chapter V, such numbers may be considered as exceptional cases and so we now shall study the solutions of $b \parallel bt \parallel < K$ for almost all $t \in [0,1)$. We begin with a result stated implicitly in F. Bernstein [1912].

Theorem 1. *Let c_1, c_2, ... be a sequence of positive numbers such that $\sum c_k$ diverges. Then the inequality*

$$B_n \parallel B_n t \parallel \ < \ c_n$$

has infinitely many solutions for almost all $t \in [0,1)$.

Proof. We shall use the methods of Theorems V.1.1 and V.1.2 to show that the set of numbers with arbitrarily long sequences of "bad" approximations has measure zero. Without loss of generality, we may suppose also that $c_k < 1$ and that $c_k \to 0$ as $k \to \infty$.

Let I_n be a given interval of order n and let $t \in I_n$ be such that $B_n \parallel B_n t \parallel > c_n$. By (I.4.3) and using (I.6.2),

$$\frac{1}{\zeta_{n+1} + \xi_n} > c_n$$

and so

$$\zeta_{n+1} < \frac{1}{c_n} - \xi_n.$$

Then, as in (V.1.3), the $t \in I_n$ such that $B_n \parallel B_n t \parallel > c_n$ form a sub-interval of

169

length

$$\frac{1}{B_n^2(1+\xi_n)} \cdot \frac{(\frac{1}{c_n}-\xi_n)-1}{(\frac{1}{c_n}-\xi_n)+\xi_n} < (1-c_n) \cdot |I_n| \qquad (1)$$

Let N be a given positive number. Let

$$S_N(0) = \{t \in [0,1) : B_N \| B_N t \| > c_N\}$$

and then, by (V.1.2) and (1), $P(S_N(0)) < 1 - c_N$. Let

$$S_N(n) = \{t \in [0,1) : B_N \| B_N t \| > c_N, \ldots, B_{N+n} \| B_{N+n} t \| > c_{N+n}\}$$

for $n = 1, 2, \ldots$. Then by induction on n we find that

$$P(S_N(n+1)) < (1-c_{N+n+1}) \cdot P(S_N(n)) < \cdots < \prod_{k=0}^{n+1} (1-c_{N+k}).$$

Since $\sum c_k$ diverges, the observation made in (V.1.4) shows that $P(S_N(n)) \to 0$ as $n \to \infty$ for any N, thus finishing the proof.

Using his estimate $\sqrt[n]{B_n} = \mathcal{O}(1)$, Khintchine [1926b] found the "dividing line" between sequences c_1, c_2, $\ldots$ of numbers having infinitely or only finitely many solutions of the approximations $n \| nt \| < c_n$ for almost all numbers.

Theorem 2. *Let $f(k)$ be a positive function defined on the positive integers such that $f(k)$ decreases monotonically to zero as k increases. Then for almost all $t \in [0,1)$ the number of solutions of $b \| bt \| < f(b)$ is (1) finite if $\sum f(k)/k$ converges and (2) infinite if $\sum f(k)/k$ diverges.*

Proof. For part (1), a solution of $b \| bt \| < f(b)$ corresponds to $|t - a/b| < f(b)/b^2$, and there are at most b such possible fractions in the interval $[0,1)$; specifically, $0/b$, $1/b$, $\ldots$, $(b-1)/b$. Around each of these fractions, there is an interval of length $2f(b)/b^2$ that could possibly contain t's with $b \| bt \| < f(b)$. Thus the measure of the solutions can be no more than

$$b \cdot \frac{2f(b)}{b^2} = \frac{2f(b)}{b}.$$

Since $\sum f(k)/k$ converges, the measure of those t's satisfying $b \,\|\, bt \,\| \, < f(b)$ infinitely many times is zero by the Borel-Cantelli lemma.

For part (2), if we can show that the divergence of $\sum f(k)/k$ implies the divergence of $\sum f(B_n)$ for almost all numbers, then Theorem 1 will give the desired result. For any given positive integer N,

$$\sum_{k=1}^{N} \frac{f(k)}{k} < \sum_{n=0}^{m} f(B_n) \left(\sum_{k=B_n}^{B_{n+1}-1} \frac{1}{k} \right),$$

where $B_m \leq N < B_{m+1}$, because f is monotonically decreasing. Since we have the asymptotic estimate

$$\sum_{k=1}^{n} \frac{1}{k} \sim \log(n),$$

the right-hand sum is asymptotically equal to

$$\sum_{n=0}^{m} f(B_n)\big(\log(B_{n+1}) - \log(B_n)\big)$$
$$= \sum_{n=1}^{m} \log(B_n) \left(f(B_{n-1}) - f(B_n) \right) + f(B_m) \log(B_{m+1}).$$

From the estimate $\sqrt[n]{B_n} = \mathcal{O}(1)$ of (V.8.1), there is a constant $K > 0$ such that $B_n < K^n$ for $n = 1, 2, \ldots$ and the above sum is

$$< K \sum_{n=1}^{m} n\big(f(B_{n-1}) - f(B_n)\big) + (m+1) \cdot K \cdot f(B_m) = K \sum_{n=0}^{m} f(B_n).$$

But $m \to \infty$ as $N \to \infty$, the divergence of $\sum f(k)/k$ implies the divergence of $\sum f(B_n)$, and we are finished.

Given this result, it is natural to try to determine for almost all numbers t the number of solutions of the n inequalities

$$k \,\|\, kt \,\| \, < f(k) \quad \text{for } 1 \leq k \leq n$$

when $\sum f(k)/k$ is divergent. As in Chapter II, we first consider relatively prime solutions and then the more general case. For each, we shall prove a strong law of large numbers in the form of an asymptotic relation because the "limiting values" will, in fact, be divergent series. However, the methods will be essentially the same as those of Chapter V.

Theorem 3. *Let $f(k) < 1/2$ be a positive function defined on the positive integers that decreases monotonically to zero as k increases and suppose that $\sum f(k)/k$ diverges. Then*

$$\sum_{\substack{|\, bt - a\, | \, < f(b)/b \\ (a,b) = 1 \\ 1 \leq b \leq n}} 1 \ \sim \ \frac{12}{\pi^2} \sum_{k=1}^{n} \frac{f(k)}{k}$$

for almost all $t \in [0,1)$.

Proof. Since $f(1) < 1/2$, we know from Chapter II that any relatively prime solution of $b\,|\,bt - a\,| < f(b)$ must be a convergent of t. We shall use the techniques of Chapter V to study the random variables

$$g_k(t) \ = \ \begin{cases} 1 & \text{if } B_k \,\|\, B_k t \,\| < f(B_k) \\ 0 & \text{otherwise} \end{cases}$$

since

$$\sum_{B_k \leq n} g_k(t) \ = \ \sum_{\substack{|\, bt - a\, | \, < f(b)/b \\ (a,b) = 1 \\ 1 \leq b \leq n}} 1 \ . \tag{2}$$

As with the proofs of Theorems 1 and V.2.2, for a given interval I_k of order k, those numbers $t \in I_k$ such that $B_k \,\|\, B_k t \,\| < f(B_k)$ form a sub-interval of length

$$\frac{1}{B_k^2(1 + \xi_k)} \cdot \frac{1 + \xi_k}{\left(\dfrac{1}{f(B_k)} - \xi_k\right) + \xi_k} \ = \ |\, I_k\,| \cdot f(B_k) \cdot (1 + \xi_k).$$

Summing over all intervals of order k, we find that the expectation of $g_k(t)$ is

$$E\big(g_k(t)\big) \ = \ \int_0^1 g_k(t)\, dt \ = \ \int_0^1 f(B_k)\,(1 + t)\, dm_k^*(t),$$

where $m_k^*(x)$ is as in Theorem V.6.1. From Theorem V.9.1, we have for any $\epsilon > 0$ that

$$f\big((\gamma+\epsilon)^k\big) \;<\; f(B_k) \;<\; f\big((\gamma-\epsilon)^k\big),$$

except for a set of measure zero, and we can make the asymptotic estimates

$$\int_0^1 g_k(t)\, dt \;=\; \frac{1}{\log(2)}\, f(\gamma^k)\big(1 \,+\, \mathcal{O}(q^k)\big)$$

and

$$E\Big(\sum_{k=1}^m g_k(t) \Big) \;\sim\; \frac{1}{\log(2)} \sum_{k=1}^m f(\gamma^k). \tag{3}$$

For the variances, we have

$$\int_0^1 \Big(\sum_{k=1}^m g_k(t) \Big)^2 dt \;-\; E^2\Big(\sum_{k=1}^m g_k(t) \Big)$$

$$= \int_0^1 \Big(\sum_{k=1}^m g_k(t) \Big) dt \;+\; 2\sum_{k=1}^{m-1}\sum_{j=1}^{m-k} \int_0^1 g_k(t)\cdot g_{k+j}(t)\, dt \;-\; E^2\Big(\sum_{k=1}^m g_k(t) \Big),$$

because $g_k^2(t) = g_k(t)$. Since the g_k's are weakly dependent, by the same methods used in Theorems V.8.1 and V.9.1, we may conclude that

$$D^2\Big(\sum_{k=1}^m g_k(t) \Big) \;=\; \mathcal{O}\Big(E\Big(\sum_{k=1}^m g_k(t) \Big) \Big).$$

In much the same manner as the proof of Theorem 2, part (2), we see that

$$\sum_{k=1}^n \frac{f(k)}{k} \;\leq\; \sum_{\gamma^k \leq n} f(\gamma^k) \Big(\sum_{\gamma^k \leq m < \gamma^{k+1}} \frac{1}{m} \Big)$$

$$= \sum_{\gamma^k \leq n} f(\gamma^k)\big(\log(\gamma) + \mathcal{O}(\tfrac{1}{\gamma^k})\big) \;=\; \frac{\pi^2}{12\log(2)} \sum_{\gamma^k \leq n} f(\gamma^k) \,+\, \mathcal{O}(1).$$

Similarly,

$$\sum_{k=1}^n \frac{f(k)}{k} \;>\; \frac{\pi^2}{12\log(2)} \sum_{\gamma^{k+1} \leq n} f(\gamma^k) \,+\, \mathcal{O}(1),$$

and we obtain the asymptotic relation

$$\frac{1}{\log(2)} \sum_{\gamma^k \leq n} f(\gamma^k) \sim \frac{12}{\pi^2} \sum_{k=1}^{n} \frac{f(k)}{k} \tag{4}$$

and that $\sum f(\gamma^k)$ is divergent.

Let

$$E(n) = E\left(\sum_{\gamma^k \leq n} g_k(t)\right).$$

By Chebyshev's inequality (V.4.1), we have

$$P\left(\left|\sum_{\gamma^k \leq n} g_k(t) - E(n)\right| > \epsilon E(n)\right) = \mathcal{O}\left(\frac{1}{\epsilon^2 E(n)}\right)$$

for any $\epsilon > 0$. Since $E(n) \to \infty$ as $n \to \infty$, there exists a subsequence $n_1, n_2, \ldots$ of the positive integers such that $\sum 1/E(n_i)$ is convergent, and on this subsequence the Borel-Cantelli lemma shows that

$$\left(\sum_{\gamma^k \leq n_i} g_k(t) - E(n_i)\right) \to 0$$

almost everywhere. Given the weak dependence of the random variables $g_k(t)$, we may repeat the proof of Theorem V.4.4 and obtain the same Kolmogorov-type inequality, but with some larger constant K instead of $4/3$. Then

$$P\left(\max_{\gamma^k \leq n_i} \left(g_1 + \cdots + g_k - E(k)\right) > \epsilon E(n_i)\right)$$

$$< K \cdot P\left(\sum_{\gamma^k \leq n_i} g_k - E(n_i) \geq \epsilon E(n_i) - \mathcal{O}\left(\sqrt{E(n_i)}\right)\right)$$

$$< K \cdot P\left(\sum_{\gamma^k \leq n_i} g_k - E(n_i) > \epsilon_2 E(n_i)\right),$$

where $\epsilon_2 > 0$ depends on ϵ and is independent of n_i. Since this last probability is $\mathcal{O}(1/(\epsilon_2^2 E(n_i)))$, by applying Chebyshev's inequality once again, the Borel-Cantelli lemma shows that the exceptional set is of measure zero and thus

$$\left(\sum_{\gamma^k \leq n} g_k(t) - E(n)\right) \to 0$$

almost everywhere. By (2), (3) and (4), this completes the proof.

We conclude with a result that is contained in a theorem of LeVeque [1960, 1959] as well as in Erdős [1959] and in Schmidt [1960].

Theorem 4. *Let $f(k) < 1/2$ be a positive function defined on the positive integers that decreases monotonically to zero as k increases and suppose that $\sum f(k)/k$ diverges. Then*

$$\sum_{\substack{b \,\|\, bt \,\| \,< f(b) \\ 1 \le b \le n}} 1 \;\sim\; 2 \sum_{k=1}^{n} \frac{f(k)}{k}$$

for almost all $t \in [0,1)$.

Proof. Since $f(1) < 1/2$, by Theorem II.5.1 the solutions of $b \,\|\, bt \,\| \,< f(b)$ must be of the form $b = cB_k$, where the positive integer c has $c^2 < f(cB_k)(\zeta_{k+1} + \xi_k)$. Since $f(B_k) \ge f(cB_k) > f(B_{k+1})$ and $0 < \xi_k < 1$, it will suffice to consider the random variables

$$f_k \;=\; f(\gamma^k)\, \zeta_{k+1},$$

as the asymptotic behavior of the f_k's is the same as that of the $f(cB_k)(\zeta_{k+1} + \xi_k)$ for almost all $t \in [0,1)$.

To find the expectation of f_k, we proceed as in (V.5.2) and find

$$E(f_k) \;=\; \frac{1}{\log(2)} \sum_{c=1}^{\infty} c \left(\log\!\Big(1 + \frac{f(\gamma^k)}{c^2}\Big) - \log\!\Big(1 + \frac{f(\gamma^k)}{(c+1)^2}\Big) \right) + \mathcal{O}\!\big(f(\gamma^k) q^k\big)$$

$$= \; \frac{1}{\log(2)} \sum_{c=1}^{\infty} \log\!\Big(1 + \frac{f(\gamma^k)}{c^2}\Big) + \mathcal{O}\!\big(f(\gamma^k) q^k\big). \tag{5}$$

Since $\sum \log(1 + f(\gamma^k)/c^2) \sim f(\gamma^k) \sum 1/c^2$ and $\sum 1/c^2 = \pi^2/6$,

$$E(n) \;=\; E\Big(\sum_{\gamma^k \le n} f_k \Big) \;\sim\; \frac{\pi^2}{6\log(2)} \sum_{\gamma^k \le n} f(\gamma^k) \;\sim\; 2 \sum_{k=1}^{n} \frac{f(k)}{k},$$

by (4).

To compute the variance $D^2(f_k)$, we must find the second moment $E(f_k^2)$, and so, in the same manner as (5), we must consider the sum

$$\frac{1}{\log(2)} \sum_{c=1}^{\infty} c^2 \left(\log\left(1 + \frac{f(\gamma^k)}{c^2}\right) - \log\left(1 + \frac{f(\gamma^k)}{(c+1)^2}\right) \right)$$

$$\sim \frac{f(\gamma^k)}{\log(2)} \sum_{c=1}^{\infty} c^2 \left(\frac{1}{c^2} - \frac{1}{(c+1)^2} \right) \sim \frac{f(\gamma^k)}{\log(2)} \sum_{c=1}^{\infty} \frac{2}{c},$$

which is divergent. As in (V.4.3), let us truncate the f_k's by setting

$$f_k^* = \begin{cases} f_k & \text{if } \zeta_{k+1} < \exp(\sqrt{E(n)}) \\ 0 & \text{otherwise} \end{cases} \quad \text{for } \gamma^k \leq n.$$

Since $\exp(\sqrt{E(n)}) \to \infty$ as $n \to \infty$, $f_k^* \to f_k$ as $n \to \infty$, $E(f_k^*) = \mathcal{O}(E(f_k))$, and

$$D^2(f_k^*) = \mathcal{O}\left(f(\gamma^k) \sum_{c=1}^{\exp(\sqrt{E(n)})} \frac{1}{c} \right) = \mathcal{O}\left(f(\gamma^k) \sqrt{E(n)} \right)$$

so that

$$D^2\left(\sum_{\gamma^k \leq n} f_k^* \right) = \mathcal{O}\left(E(n) \sqrt{E(n)} \right).$$

The remainder of the proof now proceeds as in the proof of Theorem 3, since we may use Chebyshev's inequality with $\lambda = \sqrt[4]{E(n)} \cdot \epsilon$ to find that

$$P\left(\left| \sum_{\gamma^k \leq n} f_k^* - E(n) \right| > \epsilon E(n) \right) = \mathcal{O}\left(\frac{1}{\epsilon^2 \sqrt{E(n)}} \right)$$

for any $\epsilon > 0$ and then choose a subsequence $n_1, n_2, \ldots$ of the positive integers such that $\sum 1/\sqrt{E(n_i)}$ converges.

NOTES

The estimate (V.8.1) for $\sqrt[n]{B_n}$ is sufficient for Theorem 2, but Theorem 3 requires the precise result given in (V.9.1).

The kind of problem considered in Theorem 4 originates with LeVeque [1958]; see Philipp [1971], Chapter 3, for further discussion and an extension of Szüsz [1963]. Schmidt [1964] unifies and generalizes many results of this type. See also Sprindzuk [1979].

BIBLIOGRAPHY

Adams, William W. [1966] "Asymptotic diophantine approximations to e," *Nat. Acad. Sci. USA. Proceedings.* **55**, 28-31.

Babenko, K. I. [1978] "On a problem of Gauss," *Soviet Math. Dokl.* **19** (No. 1), 136-140.

Babenko, K. I., and Jur'ev, S. P. [1978] "On the discretization of a problem of Gauss," *Soviet Math. Dokl.* **19** (No. 3), 731-735.

Barnes, E. S. [1954] "The inhomogeneous minima of binary quadratic forms. IV," *Acta Math. Acad. Sci. Hungar.* **92**, 235-264.

Barnes, E. S., and Swinnerton-Dyer, H. P. F. [1952a] "The inhomogeneous minima of binary quadratic forms. I," *Acta Math. Acad. Sci. Hungar.* **87**, 259-323. [1952b] "The inhomogeneous minima of binary quadratic forms. II," *Acta Math. Acad. Sci. Hungar.* **88**, 279-316. [1954] "The inhomogeneous minima of binary quadratic forms. III," *Acta Math. Acad. Sci. Hungar.* **92**, 199-234.

Bernstein, Felix. [1912] "Über eine Anwendung der Mengenlehre auf ein aus der Theorie der säkularen Störungen herrührendes Problem," *Math. Ann.* **71**, 417-439.

Billingsley, Patrick. [1965] *Ergodic Theory and Information.* John Wiley & Sons, Inc., New York. [1968] *Convergence of Probability Measures.* John Wiley & Sons, Inc., New York.

Borel, Émile. [1903] "Contribution à l'ánalyse arithmétique du continu," *J. Math. pures appl.* (5) **9**, 329-375.

Brillhart, John. [1972] "Note on representing a prime as a sum of two squares," *Math. of Comp.* **22**, 1011-1013.

Cassels, J. W. S. [1954] "Über $\varliminf\limits_{x \to \infty} x \,|\, \theta x + \alpha - y \,|$," *Math. Ann.* **127**, 288-304. [1972] *An Introduction to Diophantine Approximation.* Hafner Publishing Co., New York.

Cousins, Frank W. [1972] *The Solar System.* Pica Press, New York.

Cusick, Thomas W., and Flahive, Mary E. [1989] *The Markoff and Lagrange Spectra.* Math. Surveys and Monographs, No. 30. Amer. Math. Society, Providence, Rhode Island.

Davenport, Harold. [1947] "Non-homogeneous binary quadratic forms. IV," *Nederl. Akad. Wetensch., Proc.* **50**, 741-749 and 909-917 (= *Indag. Math.* **9**, 351-359 and 420-428). [1950] "On a theorem of Khintchine," *Proc. London Math. Soc.* (2) **52**, 65-80. [1982] *The Higher Arithmetic. An Introduction to the Theory of Numbers.* Fifth edition. Cambridge University Press.

Descombes, Roger. [1956a] "Sur un problème d'approximation diophantienne. I," *C. R. Acad. Sci. Paris* **242**, 1669-1672. [1956b] "Sur un problème d'approximation diophantienne. II," *C. R. Acad. Sci. Paris* **242**, 1782-1784. [1956c] "Sur la répartition des sommets d'une ligne polygonale régulière non fermée," *Ann. Sci. École Norm. Sup.* (3) **73**, 282-355.

Dickson, Leonard Eugene. [1929] *Introduction to the Theory of Numbers.* University of Chicago (reprinted 1957 by Dover, New York). [1930] *Studies in the Theory of Numbers.* University of Chicago (reprinted 1957 by Chelsea, New York). [1971] *History of the Theory of Numbers.* The Carnegie Institute of Washington (1919, 1921, 1923); reprinted by Chelsea, New York.

Dutka, Jacques. [1988] "On the Gregorian revision of the Julian calendar," *Math. Intelligencer* **10**, No. 1, 56-64.

Erdös, Paul. [1959] "Some results on diophantine approximation," *Acta Arith.* **5**, 359-369.

Forder, H. G. [1963] "A simple proof of a result on diophantine approximation," *Math. Gazette* **47**, 237-238.

Fraenkel, Aviezri S. [1985] "Systems of numeration," *Amer. Math. Monthly* **92** No. 2, 105-114.

Gauss, K. F. [1812] "Letter to Laplace of 30 January 1812," *Gauss' Werke* **X**, 1, 371-374.

Grace, J. H. [1918a] "The classification of rational approximations," *Proc. London Math. Soc.* (2) **17**, 247-258. [1918b] "Note on a Diophantine approximation," *Proc. London Math. Soc.* (2) **17**, 316-319.

Halberstam, Heini. [1986] "An obituary of Loo-keng Hua," *Math. Intelligencer* **8**, No. 4, 63-65.

Hall, Jr., Marshall. [1947] "On the sum and product of continued fractions," *Annals of Math.* **48** (No. 4), 966-993.

Hardy, G. H. [1967] *A Course in Pure Mathematics*. Cambridge University Press.

Hardy, G. H., and Wright, E. M. [1971] *An Introduction to the Theory of Numbers. 4th edition*. Oxford University Press, New York.

Hickerson, Dean R. [1973] "Length of period of simple continued fraction expansion of $\sqrt{d}$," *Pacific J. Math.* **46**, 429-432.

Hurwitz, A. [1891] "Über die angenäherte Darstellung der Irrationalzahlen durch rationale Brüche," *Math. Ann.* **39**, 279-284.

Huygens, Christiaan. [(1681)] *Projet de 1680-81, partiellement exécuté à Paris, d'un planétaire tenant compte de la variation des vitesses des planètes dan leurs orbites supposées elliptiques ou circulaires, et considération de diverses hypothèses sur cette variation* in *Oeuvres Complètes de Christiann Huygens* **21**, 109-163, Martinus Nijhoff, La Haye (1944). [(1682)] *Le planètaire de 1682* in *Oeuvres Complètes de Christiann Huygens* **21**, 165-184, Martinus Nijhoff, La Haye (1944). [(1686)] *Du livre de Wallis, Historia Algebrae Anglicè. Développement du "numerus impossibilis" (π) en une fraction continue* in *Oeuvres Complètes de Christiann Huygens* **20**, 389-394, Martinus Nijhoff, La Haye (1940). [1703] *Descriptio automati planetarii* in *Oeuvres Complètes de Christiann Huygens* **21**, 579-643, Martinus Nijhoff, La Haye (1944).

Jones, William B., and Thron, Wolfgang J. [1980] *Continued Fractions*. Addison-Wesley Publishing Company, Reading, Massachusetts.

Kac, Mark. [1959] *Statistical Independence in Probability, Analysis and Number Theory*. The Carus Mathematical Monographs, No. 12, MAA. John Wiley and Sons, New York.

Khintchine, A. Ya. [1926a] "Über eine Klasse linearer Diophantischer Approximationen," *Rendiconti di Palermo* **50**, 170-195. [1926b] "Zur metrischen Theorie der diophantischen Approximationen," *Math. Z.* **24**, 706-714. [1936] "Zur metrischen Kettenbruchtheorie," *Comp. Math.* **3**, 276-285. [1946] "On the problem of Tchebycheff," (Russian) *Izvestia Akad. Nauk SSSR* **10**, 281-294. [1963] *Continued Fractions* (trans. from 3rd Russian ed. by Peter Wynn). P. Noordhoff, Ltd., Groningen.

Koksma, J. F. [1936] *Diophantische Approximationen*. J. Springer, Berlin.

Kuzmin, R. O. [1928] "Sur un probleme de Gauss," *Atti Congr. Itern. Bologne 1928* **6**, 83-89.

Lang, Serge. [1966a] "Asymptotic diophantine approximations," *Nat. Acad. Sci. USA. Proceedings.* **55**, 31-34. [1966b] *Introduction to Diophantine Approximations*. Addison-Wesley Publishing Company, Reading, Massachusetts.

Legendre, Adrien-Marie. [1830] *Theorie des Nombres (3rd ed.)*. Reprinted 1955 by Libraire Scientifique et Technique. A. Blanchard, Paris.

Lekkerkerker, C. K. [1952] "Representations of natural numbers as a sum of Fibonacci numbers," *Simon Stevin* **29**, 190-195.

LeVeque, William J. [1953-54] "On asymmetric approximations," *Mich. Math. J.* **2**, 1-6. [1958] "On the frequency of small fractional parts in certain real sequences," *Trans. Amer. Math. Soc.* **87**, 237-261. [1959] "On the frequency of small fractional parts in certain real sequences. II," *Trans. Amer. Math. Soc.* **94**, 130-149. [1960] "On the frequency of small fractional parts in certain real sequences. III," *J. Reine Angew. Math.* **202**, 215-220. [1977] *Fundamentals of Number Theory*. Addison-Wesley Publishing Company, Reading, Mass.

Lévy, Paul. [1929] "Sur les lois de probabilité dont dépendent les quotients complets et incomplets d'une fraction continue," *Bull. Soc. Math. France* 57, 178-194. [1936] "Sur le développement en fraction continue d'un nombre choisi au hasard," *Comp. Math.* 3, 286-303. [1937] *Théorie de l'addition des variables aléatoires*. Gauthier-Villars, Paris.

Mahler, K. [1945] "A theorem of B. Segre," *Duke Math. J.* 12, 367-371.

Markowsky, George. [1992] "Misconceptions about the golden ratio," *College Math. J.* 23, No. 1, 2-19.

Mathews, Jerold. [1990] "Gear trains and continued fractions," *Amer. Math. Monthly* 97, 505-510.

Niven, Ivan. [1956] *Irrational Numbers*. The Carus Math. Monographs, No. 11, MAA. John Wiley and Sons, New York. [1963] *Diophantine Approximation*. Interscience Publishers (John Wiley and Sons), New York.

Niven, Ivan, and Zuckerman, H. S. [1980] *An Introduction to the Theory of Numbers. 4th edition*. John Wiley & Sons, Inc., New York.

Ostrowski, Alexander. [1921] "Bemerkungen zur Theorie der Diophantischen Approximationen," *Abh. Math. Sem. Hamburg. Univ.* 1, 77-98.

Perron, Oskar. [1951] *Irrationalzahlen (zweite Auflage)*. Chelsea, New York. [1954] *Die Lehre von den Kettenbrüchen. Bd. I, II. 3te Aufl.* B. G. Teubner Verlagsgesellschaft, Stuttgart.

Philipp, Walter. [1971] *Mixing sequences of random variables and probabilistic number theory*. Memoirs of AMS, No. 114. AMS, Providence, Rhode Island.

Philipp, Walter, and Stout, William. [1975] *Almost sure invariance principles for partial sums of weakly dependent random variables*. Memoirs of Amer. Math. Society, No. 161. Amer. Math. Society, Providence, Rhode Island.

Podsypanin, E. V. [1982] "Length of the period of a quadratic irrational," *Journal of Soviet Mathematics* 18, 919-923.

Rademacher, Hans. [1983] *Higher mathematics from an elementary point of view*. Birkhäuser Boston, Inc., Cambridge, Massachusetts.

Remak, Robert. [1924] "Über indefinite binäre quadratische Minimalformen," *Math. Ann.* 92, 155-182.

Rényi, Alfréd. [1970] *Probability Theory* (trans. Laszlo Vekerdi). American Elsevier, New York.

Révész, Pál. [1968] *The Laws of Large Numbers*. Academic Press, New York.

Rieger, Georg Johann. [1978] "Ein Gauss-Kuzmin-Lévy-Satz für Kettenbrüche nach nächsten Ganzen," *Manu. Math.* 24, 437-448. [1979] "Mischung und Ergodizitat bei Kettenbrüchen nach nächsten Ganzen," *J. Reine Angew. Math.* 310, 171-181.

Rockett, Andrew M. [1980] "The metrical theory of continued fractions to the nearer integer," *Acta Arith.* 38, 97-103.

Rockett, Andrew M., and Szüsz, Peter. [1986] "A localization theorem in the theory of diophantine approximation and an application to Pell's equation," *Acta Arith.* 47 (No. 4), 347-350.

Ryll-Nardzewski, Czeslaw. [1951] "On the ergodic theorem. II. Ergodic theory of continued fractions," *Studia Math.* 12, 74-79.

Schmidt, Wolfgang M. [1960] "A metrical theorem in diophantine approximation," *Canad. J. Math.* 12, 619-631. [1964] "Metrical theorems on fractional parts of sequences," *Trans. Amer. Math. Soc.* 110, 493-518.

Segre, Beniamino. [1945] "Lattice points in infinite domains, and asymmetric Diophantine approximation," *Duke Math. J.* 12, 337-365.

Series, Caroline. [1982] "Non-Euclidean geometry, continued fractions and ergodic theory," *Math. Intelligencer* 4, No. 1, 24-31. [1985] "The geometry of Markoff numbers," *Math. Intelligencer* 7, No. 3, 20-29.

Serre, Jean-Pierre. [1973] *A Course in Arithmetic*. Springer-Verlag, New York.

Shanks, Daniel. [1967] "Review of *A Table of Gaussian Primes* by L. G. Diehl & J. H. Jordan," *Math. of Comp.* **21**, 260-262.

Sós, Vera T. [1958] "On the theory of Diophantine approximations. II. Inhomogeneous problems," *Acta Math. Acad. Sci. Hungar.* **9**, 229-241.

Sprindzuk, Vladimir G. [1979] *Metric Theory of Diophantine Approximations* (trans. Richard A. Silverman). John Wiley and Sons, Inc., New York.

Szekeres, George. [1937] "On a problem of the lattice-plane," *J. London Math. Soc.* **12**, 88-93.

Szüsz, Peter. [1956] "Beweis eines zahlengeometrischen Satzes von G. Szekeres," *Acta Math. Acad. Sci. Hungar.* **7**, 75-79. [1958] "Über die metrische Theorie der diophantischen Approximation," *Acta Math. Acad. Sci. Hungar.* **9**, 179-193. [1961] "Über einen Kusminschen Satz," *Acta Math. Acad. Sci. Hungar.* **12**, 447-453. [1963] "Über die metrische Theorie der diophantischen Approximation, II," *Acta Arith.* **8**, 225-241. [1968] "On Kuzmin's theorem. II," *Duke Math. J.* **35**, 535-540. [1973] "On a theorem of Segre," *Acta Arith.* **13**, 371-377.

Tanaka, Shigeru, and Ito, Shunji. [1981] "On a family of continued-fraction transformations and their ergodic properties," *Tokyo J. Math.* **4**, 153-176.

Uspensky, J. V. [1937] *Introduction to Mathematical Probability*. McGraw-Hill, New York.

Vahlen, K. Th. [1895] "Über Näherungswerte und Kettenbrüche," *Jour. für Math.* **115**.

Wall, H. S. [1973] *Analytic Theory of Continued Fractions*. D. Van Nostrand (1948); reprinted by Chelsea, New York.

Waterman, Michael S. [1971] "A Kuzmin theorem for a class of number theoretic endomorphisms," *Acta Arith.* **19**, 31-41.

Weil, André. [1984] *Number Theory: An Approach through History; From Hammurapi to Legendre*. Birkhäuser, Boston.

Wirsing, Eduard. [1974] "On the theorem of Gauss-Kusmin-Lévy and a Frobenius-type theorem for function spaces," *Acta Arith.* **24**, 507-528.

Wölffing, E. [1908] "Wer hat über Kettenbrüche gearbeitet?" *Mathematisch-naturwissenshaftliche Mitteilungen, begründet von Dr. O. Böklen* **10** (2).

Wright, E. M. [1964] "Approximation of irrationals by rationals," *Math. Gazette* **48**, 288-289.

SYMBOLS

INDEX

A

Adams, William W., 36
approximation,
    asymmetric, 110
    by intermediate convergents, 112
    by non-convergents, 112
    homogeneous, 19
    inhomogeneous, 116, 136
auxiliary convergent, 36

B

Babenko, K. I., 166
Barnes, E. S., 136
Beppo-Levi Theorem, 141
Bernstein, Felix, 169
best approximation, 17, 19
    arc sine law, 167
    first kind, 36
    second kind, 36
Billingsley, Patrick, *viii*, 167
Borel, Émile, 135
    sets, 137
Borel-Cantelli lemma, 140
Brillhart, John, 135
Brouncker, William, 64

C

calendar construction, 60, 134
Cantor set, 72
    sum, 73
        direct construction, 135
Cassels, J. W. S., 17, 136
cattle problem of Archimedes, 134
central limit theorem, 167
Ceulen, Johannes van, 134
characteristic equation, 58
Charves, M., 41
Chebyshev's inequality, 145

Chinese remainder theorem, 51
conjugate of a quadratic surd, 43
continued fraction, 1
    complete quotient, 4
    complex partial quotient, 17
    convergent, 1
    convergent, of a reciprocal, 10
    general, 58
    partial quotient, 1
    regular, 3
    to the nearest integer, 18
    transformation, 8
    with bounded digits, 72, 74
convergent, 1
    auxiliary, 36
    denominator, 2
    intermediate, 36
    numerator, 2
    principal, 36
    quasi-, 36
Cousins, Frank W., 134
Cusick, Thomas W., 135, 136

D

Davenport, Harold, 37, 135, 136
Descombes, Roger, 37, 136
Dickson, Leonard Eugene, 17, 18, 134,
    135, 136
difference, 9
diophantine approximation, 19
    homogeneous, 19
    inhomogeneous, 116, 136
    metrical, 169
Durner, Alexander, *viii*
Dutka, Jacques, 134

E

e, 13
    approximation, 30
        by convergents, 37

185